Decision Engineering

Series Editor

Professor Rajkumar Roy
Department of Enterprise Integration
School of Industrial and Manufacturing Science
Cranfield University
Cranfield
Bedford
MK43 0AL
UK

Other titles published in this series

N. R. Milton

Knowledge Acquisition in Practice

A Step-by-step Guide

 Springer

N. R. Milton, PhD, BSc, BA
Chief Knowledge Architect
Epistemics
www.epistemics.co.uk

and

Director of Knowledge Services
Tacit Connexions
www.tacitconnexions.com

British Library Cataloguing in Publication Data
Milton, N. R.
 Knowledge acquisition in practice : a step-by-step guide. -
 (Decision engineering)
 1. Knowledge acquisition (Expert systems) 2. Knowledge
 representation (Information theory)
 I. Title
 006.3'31
ISBN-13: 9781846288609

Library of Congress Control Number: 2007925390

Decision Engineering Series ISSN 1619-5736
ISBN 978-1-84628-860-9   e-ISBN 978-1-84628-861-6   Printed on acid-free paper

9 8 7 6 5 4 3 2 1

Springer Science+Business Media
springer.com

Foreword

In the 1970s, AI practitioners came to believe that the secret to building effective software programs was to fill them with knowledge. However, they knew very little about the techniques and methods required to extract knowledge either from textural sources, databases or human experts. The field of knowledge acquisition and elicitation was born as a distinct area.

I well remember the earliest conferences and workshops on the topic. They were a heady mix of disciplines – psychologists, sociologists and computer scientists. The one unifying observation was that there was a significant bottleneck, *i.e.* that knowledge acquisition and elicitation was costly and difficult to do and there was little advice and few tools or techniques available.

As academics are wont to do in a new topic area, workshops flourished, conferences were born and research projects were funded. Significant progress was made by developing principles, methods and software tools to help extract, organise, validate and implement knowledge. A substantial problem, however, was that the area tended to become an academic enclave and too little attention was given to the actual process of engaging with end-users, application context and the business requirements. Despite 30 years of investment in this area, it is still the case that there is little practical advice and guidance for those embarking on an attempt to organise and regiment knowledge from whatever source.

During the 1980s, the discipline known as knowledge acquisition rapidly attracted the attention of not just those interested in building computer systems. The whole area of management science and organizational psychology was in need of techniques for building knowledge structures. In the late 80s and early 90s, the field of knowledge management came to prominence in a number of seminal books. Companies became increasingly interested in their knowledge assets: how they could represent the knowledge held in their companies, how they could protect it and how they could assign value to it. The whole field of knowledge management added a new dimension to the requirement for tools and techniques to formally model knowledge.

More recently, the extraordinary emergence of the largest information construct in human history, the World Wide Web, has presented us with a new way of thinking about how we might organise and regiment information. The challenge here is that a wealth of information has to be acquired, retrieved, indexed, sorted, structured and managed.

So in the last 30 years, we have seen at least 3 major drivers all of which point to a requirement for knowledge acquisition techniques. To date there has been no single treatment of the practical methods to build systems, in the most general sense, that embody knowledge. Whether these are for the purposes of building decision support systems, organizing a corporate knowledge management programme, putting up sustained intranets or modelling the content of the web.

As someone who was involved in the earliest days of the research, it gives me great pleasure to write this foreword. Nick Milton worked with me at the University of Nottingham in the early days of seeking to widen the applicability of knowledge acquisition tools and techniques. He has gained huge experience in the applied and practitioner aspects – how one teaches this material, how you convey the important concepts – and has engineered a wide variety of successful solutions for many clients.

This book presents a systematic presentation of processes, procedures and routines to organise a general knowledge acquisition project. The beauty of the approach is that it is independent of whether the project is to deliver a decision support system, a knowledge management product or a website. The approach also allows the incorporation of particular or bespoke in-house methods at various stages in the acquisition process.

As someone who on many occasions presented courses, lectures and seminars in the area, it was always somewhat embarrassing to be asked "Where is the definitive text book on the art and craft of knowledge acquisition?". This book squarely meets that requirement.

Professor Nigel Shadbolt
President of the British Computer Society
Professor of AI at the University of Southampton
January 2007

Acknowledgements

The ideas and procedure described in this book are the result of many years work by many talented people. In particular, I acknowledge the work of knowledge engineers and others at a number of organisations that have pioneered the use of knowledge engineering methods to capture, store, share and use knowledge. I especially thank those people at Airbus, BAE Systems, QinetiQ, Rolls-Royce and Thales for their contributions to this area.

I give grateful thanks to the individuals who have helped in the inception, birth and development of this book and the procedure it describes. In rough chronological order, these are: Paul Riley, Hugh Cottam, Darius Baria, Steve Swallow, Clive Emberey, Florence Sellini, David Bond, Vince Phillips, Richard McIntosh, Graeme Adamson, Graham Clarke, Spencer Wilkinson, Heather Adams, Geoff Walker, Duncan Mclean and Nigel Flood.

I give special thanks to Professor Nigel Shadbolt for his pioneering work in knowledge acquisition, for being my mentor when I first became a knowledge engineer and for writing the foreword to this book.

I give thanks to the series editor, Professor Rajkumar Roy, for his inspiring suggestion to make this book a step-by-step handbook for practicing knowledge engineers. I also thank Anthony Doyle and Simon Rees at Springer for their help and assistance.

On a personal note, I thank the four most important people in the world to me for their support and love: my wife Natasha, my daughter Margarita, and my parents Maurice and Joyce.

And last but not least, I thank you dear reader for picking up this book and reading some or all of it.

Nick Milton
January 2007

Contents

1

Introduction

Knowledge acquisition is one of those things that is easy to do badly and difficult to do well. My aim with this book is to make it easier to do well.

Let me be clear at the start what I mean by knowledge acquisition. At the heart of knowledge acquisition is the creation of a box of knowledge.

Now when I say 'box' I do not mean a real box but a computer file or group of files that hold the knowledge. Depending on who you speak to this box is variously called a knowledge base, a knowledge store, a knowledge repository or an ontology. We shall call it a knowledge base, or **k-base** for short.

By the way, there is a glossary at the end of this book in case you lose track of the jargon.

To create the k-base requires knowledge to be captured from people's heads. These people are referred to as experts, since they have a great deal of expertise in a certain field (or domain as we call it). A quick definition of knowledge is that it is equivalent to expertise. And what is expertise? It is the ability of people to do effective and efficient work and to deal with complex situations.

The final ingredient of knowledge acquisition is that we do something with the k-base, such as share it with people or computers (more of this in the next section).

Knowledge acquisition is therefore the activity of capturing expertise from people (and other sources of knowledge) and creating a computerised store of this knowledge to be used to help an organisation in some specified ways.

To perform knowledge acquisition is itself a skilled activity requiring expertise. As such there is a need for skilled people to do this. These people are called **knowledge engineers**. I am a knowledge engineer, and after reading this book I hope you will be well on the way to being one too!

1.1 Uses and Benefits

There is no point to knowledge acquisition unless we do something with the knowledge we have acquired. As it happens, there are many uses for a k-base. These fall into three main categories:

- We can use the k-base to share the knowledge with people. To do this we would normally transform the k-base into a special website called a **knowledge web**. This is uploaded to the organisation's intranet system so people can access the knowledge when they need it.
- We can use the k-base to share the knowledge with computer systems. To do this we would normally re-code the k-base into a special format called an **ontology**. This is embedded within a web system so that software systems and services can access the knowledge when they need it.
- We can use the k-base as part of the development of an intelligent computer system. To do this we would normally re-write the k-base as a document called a **knowledge document**. This is passed onto the organisation's software developers so they can develop the system, which could be an expert system, a knowledge-based system or a knowledge-based engineering system.

When performing a knowledge acquisition project, we would normally be creating a k-base for one of these uses. However, it could be re-used for another purpose, such as providing material for a process redesign, or for the auto-generation of computer code. This is an important advantage of a k-base, that it can be used and re-used for different purposes.

As you will see later in this book, a knowledge acquisition project starts by identifying an **end-product** and the **end-users** who will use the end-product. The end-product is the main deliverable of the project, *i.e.* the main thing that we produce from the k-base. The end-users are usually the people who will use or deploy the end-product. For an ontology, you could equally well think of the end-users as the software systems that will read and use the knowledge in the ontology.

How can the end-product provide specific help to the end-users? As we have seen, the end-product could be a knowledge web, an ontology or a knowledge document. Some of the benefits of these are as follows:

- A knowledge web can be used to teach people who are just starting in an area and accelerate their progress up the learning curve. This use of knowledge acquisition can be accelerated further if the new starter acts as the knowledge engineer on the project (after a suitable period of training in knowledge acquisition).
- A knowledge web can be used to spread knowledge across the functional boundaries of an organisation such as from design to manufacturing (and *vice versa*), or from technical people to financial people (and *vice versa*).
- A knowledge web can be used to archive knowledge for future generations. For example, it can store the reasons behind decisions that are made during the development of a new product. Some complex products such as military and aerospace products have an active life of many decades so it is vital that knowledge is passed down the generations in a format that is useable over a long span of time.
- A knowledge web can be used to reduce the risks involved in losing access to people who have very specific knowledge. As such, it is useful to create a knowledge web when people are close to retirement or when there are only one or two experts in a domain.

- An ontology can be used to provide a common language between software systems within a network. In this way, IT systems can use ontologies when reading, manipulating and using knowledge to perform all manner of tasks.
- An ontology can be used to give structure to informal sources of data and information, such as Wikis and Blogs. In this way information can be fused and filtered to provide people with better views of the information they require.
- An expert system can provide people with advice by replacing part of the reasoning that is performed by experts. In fact, experts can use such a system themselves to reduce workload when there is too much to do and too little time.
- An expert system can enable inexperienced people to perform complex activities by providing suggestions and advice.
- A knowledge-based engineering system can perform certain knowledge-intensive tasks such as designing complex components in a fraction of the time it takes a human to do the task.

As can be seen from these examples, knowledge acquisition has an important role to play in a number of fields such as knowledge management, knowledge engineering, knowledge-based engineering and ontological engineering. If you wish to read more about these areas I suggest you take a look at some of the resources shown in the references and bibliography at the end of the book.

1.2 The Nature of Knowledge

I have been talking about knowledge and have briefly described it as being equivalent to expertise. Let me say a little more about this and introduce you to some important types of knowledge.

1.2.1 What is Knowledge?

No decent book in the areas of knowledge engineering and knowledge management fails to ask the question: What is knowledge? The usual answers are variations on a number of themes:

- Knowledge is a highly-structured form of information;
- Knowledge is what is needed to think like an expert;
- Knowledge is what separates experts from non-experts;
- Knowledge is what is required to perform complex tasks

All of these are useful ways of thinking about knowledge. Another way is to think in terms of this: Knowledge is a like a machine or an engine in our heads.

What do I mean by this? I mean that knowledge is an active thing that manipulates, transforms or creates something out of something else. It is a machine in someone's head that takes in data and information at one end and spurts out decisions and actions at the other end. Let me put this idea into a definition:

$$\text{Knowledge is the} \left\{ \begin{array}{c} \text{ability} \\ \text{skill} \\ \text{expertise} \end{array} \right\} \text{to} \left\{ \begin{array}{c} \text{manipulate} \\ \text{transform} \\ \text{create} \end{array} \right\} \left\{ \begin{array}{c} \text{data} \\ \text{information} \\ \text{ideas} \end{array} \right\} \text{to} \left\{ \begin{array}{c} \text{perform skilfully} \\ \text{make decisions} \\ \text{solve problems} \end{array} \right\}$$

Just like a real machine, the knowledge machine in a person's head can only be completely understood if you know two things:

- How it is **structured**, *i.e.* what components it is made from, and the ways they are linked together;
- How it **operates**, *i.e.* the ways in which the components behave and the processes that are happening

As we shall see below, this is an important way in which we can view knowledge, as being about its structural components or about the processes that operate.

1.2.2 What Types of Knowledge Are There?

There are various ways of looking at knowledge and describing it as being one thing or another. Two important dimensions with which to describe knowledge are:

- Procedural knowledge *vs*. Conceptual knowledge;
- Basic, explicit knowledge *vs*. Deep, tacit knowledge

1.2.2.1 Procedural Knowledge

Procedural knowledge is the knowledge you would name if you were completing the sentence: "I know how to…". So if I say: "I know how to drive a car" then this type of knowledge is procedural. Hence, it is about processes, tasks and activities. It is about the conditions under which specific tasks are performed and the order in which tasks are performed. It is about the resources required to perform tasks and it is about the sub-tasks that are required.

1.2.2.2 Conceptual Knowledge

Conceptual knowledge is what you would say if you were completing the sentence: "I know that…". So if I say: "I know that a sports car goes faster than a lorry" then this type of knowledge is conceptual. Hence, it is about the ways in which things (which we call 'concepts') are related to one another and about their properties. An important form of conceptual knowledge concerns taxonomies, *i.e.* classes and class membership. Hence, taxonomic knowledge is about answering the question: "What sort of thing is a…?". For example: What sort of thing is a dolphin? (We shall see in Chapter 2 that it is a type of whale!). Another type of conceptual knowledge is about the attributes of concepts. I will cover all about relationships and attributes in Section 2.2 in Chapter 2.

1.2.2.3 Basic, Explicit Knowledge

As the name suggests this type of knowledge is at the forefront of an expert's brain and is thought about in a deliberate and conscious way. It is concerned with basic tasks that an expert performs, basic relationships between concepts, and basic properties of concepts. This type of knowledge is not too difficult to explain and is the sort of thing that is taught in classrooms and lecture theatres.

1.2.2.4 Deep, Tacit Knowledge

This is at the other extreme to basic, explicit knowledge. It is knowledge that is thought about at the back of one's brain, in what some people call the 'subconscious'. It is often built up from experiences rather than being taught. Hence, it is the sort of knowledge that someone gains when they practice something. It often leads to automatic activities that seem to require no thought at all (at least no conscious thought). It is described in everyday words and phrases such as 'gut feel', 'hunches', 'intuition', 'instinct' and 'inspiration'. An example of tacit knowledge is driving a car, which experienced drivers can do 'without thinking' but learner drives find difficult because they lack the deep, tacit knowledge.

1.2.2.5 Examples

Some knowledge engineers tend to think that knowledge must be one thing or another. So it is either procedural or it is conceptual; it is either explicit or it is tacit. In practice, things are less black and white and more shades of grey. Knowledge types lay along a continuum. For example, you may know something (*e.g.* someone's name) but not be able to articulate it because it has 'slipped your mind'. In other words, you cannot say it but it is somewhere in your memory. When this happens, a simple cue, such as the initial letters of the words, can help you to recall the words. Thus we can think of knowledge as laying somewhere on a dimension between pure tacit knowledge and pure explicit knowledge, and the ease at which it can be recalled will determine its position along this dimension.

To my mind the best way to picture different types of knowledge is as a graph of 2 dimensions, so that any particular piece of knowledge can be placed somewhere in the space. Figure 1.1 shows such a graph with some examples.

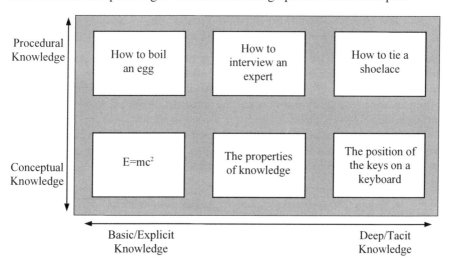

Figure 1.1. Examples of different types of knowledge using a 2-dimensional graph

Let me provide a brief explanation of Figure 1.1. 'How to boil an egg' is placed in the top-left as it is a simple task that can easily be explained. 'How to tie a shoelace' is in the top-right because it is a task that is almost impossible to explain without doing it and commentating as you do it. 'E=mc^2' is placed in the bottom-left as it is a basic relationship between concepts. 'The position of the keys on a keyboard' is something many people know subconsciously (*i.e.* those who regularly type into computers) but very few know consciously (can you say which keys are to the left and right of the letter B?).

We shall see this 2-dimensional space of knowledge again in Chapter 2 when I compare different capture techniques (see Figure 2.1 on page 11).

1.3 Issues and Difficulties

We have seen that knowledge acquisition is about creating a store of knowledge that can be used to provide an end-product that has many different applications and benefits. We have examined what knowledge is and some different types. Now let me tie these areas together by starting to take a look at the issues and difficulties of acquiring knowledge.

1.3.1 Knowledge Acquisition Issues

There are a number of important factors that we must bear in mind when running a knowledge acquisition project:

- The end-product must be useful to the end-users;
- To be useful, the end-product must be full of high-quality knowledge that is correct, complete and relevant and that is stored in a structured manner;
- The project must be run in an efficient way making the most use of the available resources;
- The project should not unduly disrupt the normal running of the organisation, hence should not involve too much time from experts

Now I hope you can see the real issue at the heart of knowledge acquisition and why a book of this kind is required. We need to do something quite difficult (extract a large mass of knowledge from deep inside people's heads) but do so as efficiently as possible. In other words, we must maximise the knowledge we get whilst minimising the resources used.

This is not easy, especially when you need to get at deep, tacit knowledge or you have experts who disagree with one another. The key to all this is to base what we do on a number of tried and tested principles:

- We must be **systematic**. In other words, we must have methodologies to follow and frameworks to use. These help us to focus on the knowledge that is needed and obtain it in the most efficient manner.
- We must, whenever possible, **re-use** knowledge from previous projects or from generic models of knowledge, such as generic taxonomies. In this

way we need not start with a blank k-base, but instead have a skeleton to fill-in that provides prompts for the sorts of knowledge to capture.

- We must have at our disposal a **range of techniques** to use when acquiring, analysing and modelling knowledge. In other words, we need a toolkit from which we can select the right tool for the job. As we have seen, there are different types of knowledge, so we need different techniques for dealing with them.
- We must, whenever possible, make use of special **software** that can help make our jobs easier, quicker and more effective.
- We must **learn from other people** and share around our own expertise of knowledge acquisition.

You will see in Chapter 2 how I expand on these basic principles and show how we realise them in three key areas: knowledge capture, knowledge analysis and knowledge modelling. You will also see how the 47-step procedure that is described in this book, has these principles as its very heart. In this way, it is similar to a number of other methodologies, notably CommonKADS and MOKA.

The 47-step procedure in this book has many similarities with methodologies like CommonKADS and MOKA and can be used alongside them to create more specific methods. The main difference between the 47-step procedure and other methodologies is its generality, *i.e.* the 47-step procedure can be used on any knowledge acquisition project whether it is on bee keeping, aircraft design, corporate law or military planning; whether the end-product is a website, an ontology or an intelligent software system; whether the knowledge engineer has had 2 days of training or has a doctorate in artificial intelligence.

1.3.2 Difficulties of Capturing Knowledge

Let me conclude this introduction to knowledge acquisition with some of the specific difficulties we encounter when we meet with experts to capture their knowledge. These difficulties come from two sides: difficulties from the expert's side and difficulties from the knowledge engineer's side. Experts can find it difficult to:

- Express their expertise in a manner that is fully comprehensible to the knowledge engineer;
- Ascertain what the knowledge engineer actually wants;
- Give the right level of detail;
- Present ideas in a clear and logical order;
- Explain all of the jargon and the domain-specific terminology;
- Recall everything that is relevant to the project;
- Avoid drifting off to talk about irrelevant things

Knowledge engineers can find it difficult to:

- Understand everything the expert says;
- Note down everything the expert says;
- Keep the expert talking about relevant issues;

- Maintain the high level of concentration required to take in a mass of new knowledge;
- Check that they have fully understood what was said

These difficulties arise because of the way the human mind works. We are good at communicating – evolution has seen to that – and we are good at developing expertise in performing complex activities. But we are not so good at communicating complex ideas with people who are not in the same subject area. These problems stem from our psychology and our communicative limitations. Did I sit down and write my own expertise into this book without many re-writes and struggles? I did not. I worked hard for many months to put into words the stuff in my head that I use every day 'without thinking'.

   We shall see in the following chapters many methods that knowledge engineers have developed to deal with these difficulties. Using these methods ensures that a knowledge acquisition project can capture masses of high-quality knowledge in an efficient way.

1.4 About this Book

1.4.1 How this Book is Structured

The rest of this book is structured as follows. The next chapter introduces a procedure of 47 steps that covers the activities required for a successful knowledge acquisition project. Chapters 3-6 give detailed information of this procedure. Each of these chapters presents one of the four main phases of the 47-step procedure. The book concludes with a chapter that provides supplementary advice on implementation issues, such as managing a knowledge programme and supporting knowledge engineers.

1.4.2 What This Book Can Do For You

This book has four main uses:

- If you have a knowledge acquisition project to perform and you need to know what to do, then the step-by-step procedure in this book will be guide you through every stage of the project;
- If your organisation uses a standard methodology, such as CommonKADS or MOKA, and you wish to enhance it with some different techniques, then the step-by-step procedure in this book will give you some options;
- If you are running, or plan on running, a programme in your organisation to capture, manage and use knowledge, then this book will give you many ideas on the issues, resources and methods involved;
- If you are interested in knowledge acquisition and knowledge engineering and wish to know more about the practical side of these areas, then this book will provide this information.

2

Overview of the Step-by-step Procedure

This chapter introduces a procedure of 47 steps that covers the activities required for a successful knowledge acquisition project. I start by introducing three essential aspects of the procedure: knowledge capture (techniques to use when seeing experts), knowledge analysis (identifying the elements required to build the k-base) and knowledge modelling (creating different ways of editing and viewing the k-base). I will then walk you through a complete project that uses the 47-step procedure. The chapter ends with a list showing how the steps in the procedure contribute to the usefulness, usability and use of the end-product.

2.1 Knowledge Capture

2.1.1 Knowledge Capture Techniques

Key parts of a knowledge acquisition project are the encounters with experts. During the stages that capture domain knowledge, these sessions are concerned with two things: (i) **Eliciting** knowledge, *i.e.* capturing knowledge that is not in the k-base; (ii) **Validating** knowledge, *i.e.* checking that knowledge in the k-base is correct, complete and relevant to the project

What techniques can be used for eliciting and validating knowledge? As introduced in Chapter 1, we need to do more than ask an expert a few questions and expect knowledge to pour forth. We need to deal with memory effects and communication difficulties. We need to deal with our own lack of understanding. We need to deal with experts having different opinions.

To do this requires having a range of techniques that are based on a clear understanding of the psychology of expertise. Such a body of techniques has been developed over the past 25 years by knowledge engineers, and includes techniques that have undergone critical analysis using controlled experiments (Shadbolt, 2005). Emerging from this work are three groups of techniques: interview techniques, modelling techniques and specialised techniques.

2.1.1.1 Interview Techniques

Interview techniques involve questioning the experts. They are good for eliciting basic knowledge but not good for validating knowledge. There are 3 variants: unstructured interview, semi-structured interview and structured interview.

- The unstructured interview has very little planning and is a freeform chat with the expert. This can be used in the early stages of elicitation to get some basic knowledge of the domain but is not normally used for most elicitation sessions, as it is not very efficient.
- The semi-structured interview is the main technique for eliciting explicit knowledge. It uses a pre-defined set of questions that are sent to the expert beforehand, and supplementary questions that are asked at the interview. Steps 15 and 16 of the 47-step procedure are specifically concerned with preparing and conducting semi-structured interviews.
- The structured interview uses a pre-defined set of questions and no supplementary questions. It often involves a questionnaire that is filled-in at the session. This is usually preferable to sending questionnaires to people, as they rarely respond to them. The structured interview has its uses but it is not a main technique used in the 47-step procedure.

2.1.1.2 Modelling Techniques

Modelling techniques involve the use of knowledge models with experts. Knowledge models, or **k-models** for short, are ways of viewing the k-base using different forms of diagrams and matrices (see Section 2.3 for a full description and many examples). K-models can be used in three ways:

- A rudimentary k-model is shown to an expert and he/she is prompted to modify the errors and fill-in the gaps. This is used in Step 29 of the procedure to validate the knowledge from interviews and to prompt the expert to add more detailed knowledge.
- A complete k-model is shown to the expert who provided the knowledge or a secondary expert. The latter is the main way to perform cross-validation in Step 36 of the procedure.
- A k-model is created from scratch on a blank sheet of paper or blank computer screen. This technique can be used to capture basic knowledge (*e.g.* use of a timeline in Step 16) or to elicit deeper knowledge (*e.g.* the use of a concept map in Steps 30 and 35).

2.1.1.3 Specialised Techniques

Specialised techniques (a.k.a. contrived techniques) involve methods developed by psychologists for probing the expert's mind for deep, tacit knowledge. They tend to involve the expert performing a task that will expose and reveal knowledge that is difficult to articulate. These tasks come in two varieties: (i) Tasks that the expert normally performs, such as in the commentary technique; (ii) Tasks that have been devised to probe the expert, such as in the concept sorting and triadic elicitation techniques. A description of the main specialised techniques can be found in Step 35 (on pages 125-132).

2.1.2 Comparing the Capture Techniques

Having briefly surveyed the main classes of capture techniques let me now turn to some of the similarities and differences between them.

2.1.2.1 Similarities Between the Capture Techniques
All of the capture techniques do three main things:

1. They **focus** the expert on the knowledge that is required, and discourage the expert from wandering away from that;
2. They help the expert to **recall** knowledge by providing prompts or by approaching the knowledge from different angles;
3. They help the expert to **explain** what he/she knows in a clear way

2.1.2.2 Differences Between the Capture Techniques
One of the main differences between the techniques is that interview techniques and specialised techniques are used to elicit knowledge but not to validate knowledge whereas modelling techniques can do both. Another difference is the type of knowledge that each technique is best at eliciting. Using the two main dimensions of knowledge described in Chapter 1, Figure 2.1 shows which techniques are most appropriate to use for particular types of knowledge.

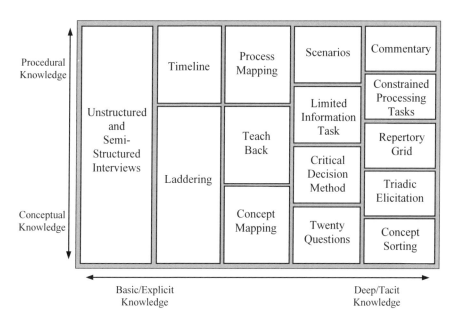

Figure 2.1. Techniques for capturing different types of knowledge

In Figure 2.1, interview techniques are shown on the left indicating that they are most effective for basic, explicit knowledge that is both conceptual and procedural. Modelling techniques (timeline, laddering, process mapping, teach back and concept mapping) are shown in the centre of the horizontal axis as they are used to capture more detailed knowledge than interviews. Timeline and process mapping are at the top indicating that they are most useful for procedural knowledge, and concept mapping and laddering are towards the bottom showing that these are more effective for conceptual knowledge. Specialised techniques are shown on the right as they are used to capture deep, tacit knowledge. An explanation of each of the specialised techniques is contained in Step 35 (see Section 5.6).

2.2 Knowledge Analysis

Knowledge analysis is an activity performed by the knowledge engineer after he/she has had a knowledge elicitation session with an expert. It is concerned with identifying elements of knowledge that will be entered into the k-base to form its structure and main components. These elements are those that will be used as building blocks to form all of the k-models. There are four important elements (a.k.a. knowledge objects) that can be identified during knowledge analysis: concepts, attributes, values and relations. Let us take a look at each of these.

2.2.1 Concepts

Concepts are the things that constitute a domain. Some of the main types are:

- Physical concepts, *e.g.* products, components, machines;
- Pieces of information, *e.g.* ideas, plans, goals, requirements;
- Sources of information, *e.g.* documents, databases, websites;
- People and roles, *e.g.* domain experts, roles of domain experts;
- Organisations and groups, *e.g.* producers, suppliers, divisions;
- Areas of knowledge, *e.g.* Physics, Corporate Law, Systems Analysis;
- Functions, *e.g.* the purpose of components and roles;
- Tasks, *e.g.* the activities performed by experts;
- Issues, *e.g.* problems, solutions, advantages, disadvantages;
- Physical phenomena, *e.g.* physical mechanisms and forces;
- Others, *e.g.* materials, behaviours, constraints, states

Concepts form the main structure of the k-base. The other elements of the k-base are there to describe the concepts. This is done in two ways: (i) The properties of concepts are described using **attributes** and **values**; (ii) The ways in which pairs of concepts relate to one another are described using **relations**.

2.2.2 Attributes

Attributes are the qualities or features belonging to a class of concepts. In other words, they are the ways in which we see concepts as being different from each other. Some examples of attributes are:

- Attributes of physical objects, *e.g.* weight, shape, age;
- Attributes of information, *e.g.* source, format, importance;
- Attributes of people, *e.g.* gender, age, salary, personality;
- Attributes of organisations, *e.g.* size, turnover, product range

2.2.3 Values

Values are the specific qualities or features belonging to a concept that distinguish it from other concepts. Each value is always associated with an attribute. Some examples of values and their associated attributes are:

- The values *red, green* and *black* are associated with the attribute *colour*;
- The values *light* and *very heavy* are associated with the attribute *weight*;
- The values *0, 1* and *4* are associated with the attribute *number of children*;
- The values *3.7* and *8.2* are associated with the attribute *acid-test ratio*;
- The values *4-legged marsupial* and *Technique for calculating the roots of a quadratic equation* are associated with the attribute *glossary entry*

As can be seen in these examples, values come in different varieties. Some values are adjectives, some are numbers and some are sentences. Others can be paragraphs of text or chunks of hypertext code that include hyperlinks, images and pictures. To reflect these varieties, we often group attributes into classes, such as:

- Categorical attributes for values that are adjectives, *e.g. red* and *heavy*;
- Numerical attributes for values that are numbers, *e.g. 1* and *8.2*;
- Text attributes for values that are one or two sentences;
- Hypertext attributes for values that are chunks of hypertext;

2.2.4 Relations

Relations are the ways in which pairs of concepts are associated with one another. For example, what is the relation between 'cheese' and 'food'? Obviously it is that cheese **is a type of** food. This type of relation tells us what class something belongs to, and is usually shortened to '**is a**'. This is the most important relation. Other important relations are:

- 'has part', *e.g.* car – has part – engine;
- 'performs', *e.g.* knowledge engineer – performs – create concept tree;
- 'followed by', *e.g.* create model – followed by – validate model;
- 'requires', *e.g.* electric torch – requires – batteries;
- 'causes' , *e.g.* contact with acid – causes – corrosion;
- 'produces', *e.g.* KA project – produces – knowledge web

Every relation can also have an 'inverse' that goes in the opposite direction. For example, the inverse of 'has part' is 'part of', and the inverse of 'performs' is 'performed by'. Some k-bases will include the inverses of some relations and others will not. It depends on how you want to store and model the knowledge.

To clarify some terminology: when a relation is connecting two concepts, then it is referred to as a 'relationship' or a 'triple' (because there are 3 elements in it).

2.3 Knowledge Modelling

Knowledge modelling involves the creation and use of knowledge models (or **k-models** for short). K-models are ways of viewing the knowledge contained in a k-base. The k-base is a confusing mass of interconnected knowledge, so it is important that we have ways of viewing parts of it using simple and clear views. Each k-model should be thought of as providing a different perspective or different viewpoint from which to see different aspects of the k-base.

As mentioned previously, k-models are vital for helping to elicit knowledge from experts and for validating the knowledge that has been captured. K-models are also used in the end-product to convey knowledge to end-users.

Some of the main k-models are trees, matrices, maps, timelines, frames and k-pages. Information and examples of these are given in the following sections.

2.3.1 Trees

A tree is a diagram that shows a hierarchical arrangement of nodes. Each node represents a concept in the k-base and each link represents a relationship between a pair of concepts. Some important types are concept tree, composition tree, process tree, attribute tree, cause tree and mixed tree. These are described below.

A **concept tree** is the most important tree, and possibly the most important k-model. It is characterised by having every link as the 'is a' relation. Hence, it shows what type of thing everything in the k-base is. This form of knowledge is called a taxonomy and is a key aspect of the k-base. An example of a concept tree is shown in Figure 2.2 (where we see that dolphins are a type of whale).

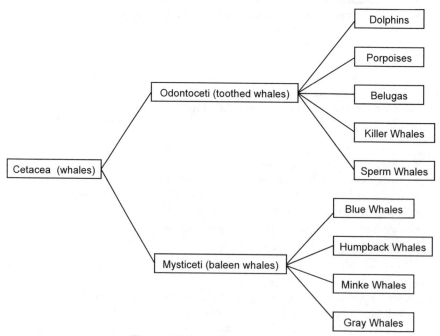

Figure 2.2. Example of a concept tree

A **composition tree** is characterised by having all the links as the 'has part' relation. This is used to show the components and sub-components of a concept such as a complex product, a document or an organisation. An example of a composition tree is shown in Figure 4.8 (on page 99).

A **process tree** is a special form of composition tree in which all the nodes are tasks. Hence, it shows how a complex task is composed of sub-tasks and sub-sub-tasks, *etc*. An example is shown in Figure 4.10 (on page 106).

An **attribute tree** shows the attributes and values that describe the properties of certain concepts in the k-base. An example is shown in Figure 4.9 (on page 102).

A **cause tree** is characterised by having the 'caused by' relationship for all links. It is used in projects that require knowledge to be captured of how experts diagnose a problem.

A **mixed tree** is a tree that contains more than one type of relation. An example is shown in Figure 2.3. The relations shown on the tree in Figure 2.3 are 'is a' for the black lines, 'has exhibit' for the black arrowed lines and 'painted by' for the dashed lines.

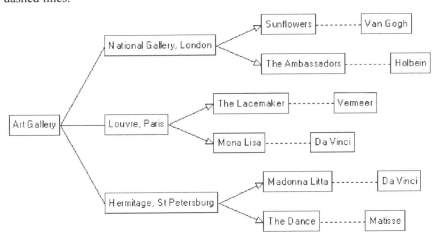

Figure 2.3. Example of a mixed tree

Note by convention 'is a' is read from right to left (*i.e.* is directed towards the highest, root node of a tree), whereas all other relations are read from left to right (*i.e.* directed away from the root node).

2.3.2 Matrices

There are two main types of matrix, an attribute matrix and a relationship matrix.

An **attribute matrix** is a way of showing the properties (attributes and values) of a set of concepts. It does this by presenting a matrix with concepts on the vertical axis and attributes/values along the horizontal axis. An example of an attribute matrix is shown in Figure 2.4, which shows the properties of a set of drinks.

		transparency			alcohol level		serving temp		fizziness	
		opaque	semi opaque	transparent	alcoholic	non-alcoholic	served hot	served cold	fizzy	not fizzy
drinks	lemonade			▬		▬		▬	▬	
	water			▬		▬		▬		▬
	vodka			▬	▬			▬		▬
	lager		▬		▬			▬	▬	
	coffee	▬				▬	▬			▬

Figure 2.4. Example of an attribute matrix

A **relationship matrix** shows two sets of concepts related to one another using a specified relation. The cells in the matrix show which pairs of concepts have the specified relationship. Figure 2.5 shows an example of a relationship matrix. This matrix shows people linked to areas of knowledge using the 'has expertise' relation, *e.g.* 'Jane Lenton – has expertise – financial management'.

		people							
		Jane Lenton	Tom Mansfield	Ian Chesterfield	Dave Newark	Natasha Dacha	Maurice Leeds	Simon Bradford	Carrie Lincoln
areas of knowledge	financial management	☐					☐		
	resource management		☐			☐			☐
	outsourcing	☐	☐	☐				☐	
	business transformation				☐			☐	
	business analysis	☐	☐	☐			☐		

Figure 2.5. Example of a relationship matrix

2.3.3 Maps

A map is a diagram that shows an arrangement of nodes linked by arrows. Each node represents a concept in the k-base and each link represents a relationship between a pair of concepts. Hence, a map is like a mixed tree that need not be hierarchical. The most important types are concept maps and process maps (see below). Other diagram formats can also be used, such state diagrams (see Figure 5.1 on page 132).

A **concept map** shows a variety of concepts connected by a mixture of different relations. An example is shown in Figure 2.6. Concept maps can come in different varieties, such as hierarchical concept maps (similar to mixed trees) and those that restrict the concepts and relations that are shown. Other names for a concept map are conceptual map, entity-relation diagram and semantic network.

The use of concept maps has been advocated as a comprehensive technique for capturing knowledge from experts (Zaff *et al.*, 1993; Cañas *et al.*, 2004).

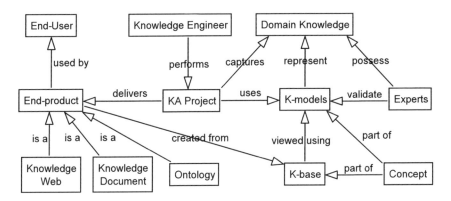

Figure 2.6. Example of a concept map

A **process map** shows the way a task (process, activity) is performed. The main elements on a process map are the sub-tasks of the task that is being modelled. These sub-tasks are placed on the map in the order in which they are performed. Links between tasks represent the 'followed by' relation. Other concepts can be shown on the process map with links to the task nodes using arrows. These can include: the resources required to perform each task; the products of each task; the triggers that cause a task to start; the people, roles or things that perform each task; the decision points that affect which tasks are performed. An example of a process map is shown on Figure 4.11 (on page 107).

2.3.4 Timeline

A timeline is a diagram that shows time along the horizontal axis and contains concepts as nodes. The width of each node shows when the concept starts and finishes. This can be used to show the phases of a project or the order of events or tasks. It is a simple representation that is often used in the early stages of knowledge elicitation to capture the basics of processes from the expert (see Steps 15 and 16). An example of a timeline is shown in Figure 2.7.

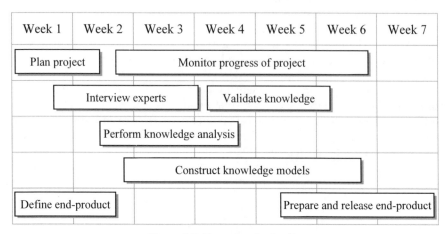

Figure 2.7. Example of a timeline

2.3.5 Frame

A frame is a simple 2-column table that shows the properties (attributes and values) of a concept. It is arranged so that the attributes associated with the concept are shown in the left-hand column and the corresponding values are shown in the right-hand column. Table 4.3 (on page 101) shows two frames, one showing the attributes and values of coffee the other showing the attributes and values of vodka.

2.3.6 K-page

A k-page is a simple 2-column table that shows all of the knowledge associated with a concept. It is very similar to a frame but shows more information. In addition to attributes and values, it shows relationships to other concepts and less formal knowledge such as descriptive paragraphs of text, pictures and images. If created using a web tool, it includes hyperlinks to other concepts in the k-base and to other files outside the k-base (such as documents, image files and video files). Figure 4.12 (on page 108) shows an example of a k-page.

2.3.7 Advanced Modelling

We have seen in these sections on k-models that there are various ways of showing the relationships between pairs of concepts or the properties of concepts. Some projects require more advanced modelling. Let us take a quick look at three advanced modelling techniques.

2.3.7.1 Properties of Relationships

Sometimes we want to show that a relationship (triple) has properties. For example, suppose I have the relationship 'Joe Smith – works for – Epistemics'. I may want to say how many years Joe Smith has worked for Epistemics. In this case, I want an attribute 'number of years' associated with the relationship. Some k-base technologies can handle this and others cannot. If they can, we need a way of showing this. One way is to use a relationship matrix but instead of simple square symbols in the cells, we show the values of the attribute (as in Figure 4.5 on page 93). Another way would be to have a frame or k-page for the relationship 'Joe Smith – works for – Epistemics'.

Properties of relationships are useful when modelling procedural knowledge. We saw earlier in Section 2.3.3 that a process map can include decision points that show the conditions under which the tasks that follow will occur. These conditions are usually shown as labels on the arrows from the decision point to the task. An example of this is on Figure 4.11 (page 107) where the labels 'Yes' and 'No' are used after the 'JICs Required?' node. To model this in the k-base requires an attribute of 'condition' on the 'followed by' relation, with values 'Yes' and 'No'.

2.3.7.2 Relationships Between Relationships
On rare occasions we need to show a relationship between a concept and a relationship, or between two relationships. To do this, requires treating one or more relationships as concepts. This is called 'reification' which means making real. For example, 'Joe Smith – works for – Epistemics' could be reified as a concept so it can have relationships with concepts or other reified relationships. An example of this might be to model the fact that Joe Smith works for Epistemics in their Nottingham office.

2.3.7.3 Rules and How to Model Them
Rules are statements of the form 'If… then…'. For example: 'If the room is on fire then activate the fire alarm', or 'If the animal has wings and feathers then it is a bird'. These statements contain one or more concepts in complex combinations with other concepts or attributes and values. Hence, we require a complex form of modelling. One way is to split the antecedent (the part before 'then') and the consequent (the part after 'then'). The consequent may already be a concept such as a task. If not, then we can model it using one of the techniques described above. Finally, the relationship between the antecedent and consequent can be a simple one such as 'implies'. Note if we want to keep things simple, we can just enter the rule into a k-page as a piece of text rather than formally modelling it.

2.4 47-step Walkthrough

We have seen in the previous sections three important aspects of a KA project: knowledge capture, knowledge analysis and knowledge modelling. In particular, we have seen many ways of capturing and displaying knowledge. Let us now see how these elements fit together into a project. This will introduce you to each of the 47 steps in the procedure described in the next four chapters.

To do this I will take you on a very brisk walk through a typical project to show what happens at each stage. Details of each step and variations from the norm are described in detail in Chapters 3 to 6. During the walkthrough, I will assume you are the knowledge engineer conducting the project.

2.4.1 Phase 1: Start, Scope and Plan the Project

The starting point for any project is to identify a project idea. This includes how the project can benefit the organisation and what the project would involve (Step 1). Opinions on the project idea are gathered from all relevant people (Step 2) and the idea is then documented as a 'project proposal' (Step 3). Agreement is sought for the project proposal from all key people (Step 4).

The initial phase of knowledge capture can now begin in order to scope the project, *i.e.* define which specific areas of knowledge that will be acquired. To start this, you (the knowledge engineer) creates a k-base (Step 5) so you have a place to put the knowledge. With the involvement of domain experts, you break the domain into different topics (Step 6) and then rate or rank these topics against key criteria (Step 7). Using this information, you identify a proposed scope and finalise it with the relevant people (Step 8).

You can now identify the sources of knowledge that will be needed during the project (Step 9). It is a good idea to identify exactly what sort of project you are doing (Step 10) which will help to define and understand the procedure that you will follow for the remainder of the project (Step 11). The procedure is used to create a project schedule (Step 12). The final part of the first phase is to collate the project proposal, project scope and project schedule into a project plan and disseminate this, and any other materials, to the project team (Step 13).

2.4.2 Phase 2: Initial Capture and Modelling

The starting point for Phase 2 is to learn the basics of the domain from documents or informal conversations with domain experts (Step 14). Following this, you prepare for the first one or two semi-structured interviews (Step 15) and then conduct these interviews (Step 16). The audio-recordings of these interviews are transcribed, either fully or partially (Step 17).

Using the interview transcripts and domain documents, you analyse the knowledge to identify the concepts that will form the main structure of the k-base (Step 18). A concept tree can now be created to develop a taxonomy of the concepts (Step 19). The concept tree is validated with the domain experts and added to where necessary (Step 20).

At this point, it is useful to take stock of the project and update the plan if necessary in light of what you have learnt (Step 21).

A k-page describing each concept is created (Step 22), and a glossary of all domain-specific terms is begun (Step 23). A meta-model is defined that shows how the k-base will be structured in terms of the relationships between concepts and the properties of concepts (Step 24). The meta-model is used to set-up the structure of the k-base and to define the templates for k-models (Step 25).

The relationships between concepts are entered using the appropriate k-models (Step 26). The properties (attributes and values) of certain concepts can be entered if required (Step 27). Knowledge of processes and how the expert performs certain tasks is entered into the k-base (Step 28).

Phase 3 ends with one or more validation sessions with the experts. A number of k-models are used to validate that the knowledge is correct, complete and relevant to the project (Step 29). A first-pass k-base is now in place, which for some small projects may be all that is required.

2.4.3 Phase 3: Detailed Capture and Modelling

Further interviews and modelling activities are used to capture more detailed knowledge (Step 30). The scale of the project and the requirements of the end-product will determine the extent of this step.

The main k-models can now be finalised (Step 31) in preparation for the creation of a prototype end-product (Step 32). This prototype is used to carry out an assessment exercise with a representative sample of end-users (Step 33). Based on the results of the assessment exercise, a completion plan is produced that defines the actions required to complete the project (Step 34).

In light of feedback from end-users and the project scope, the capture of very detailed (deep, tacit) knowledge now takes place using a suitable set of specialised techniques (Step 35). If required, cross-validation takes place with secondary experts checking that the k-models developed with the primary experts are correct and best practice (Step 36). Any differences of opinion between experts are identified, discussed and resolved at a consensus session (Step 37).

Before finalising the k-base, a check is made to ensure that all of the knowledge has been validated and its status has been recorded (Step 38). The k-models and the underlying k-base can now be finalised (Step 39).

2.4.4 Phase 4: Share the Store Knowledge

The final phase starts by defining and creating the format of the end-product (Step 40). Using the k-base and the format, a provisional end-product is created (Step 41). This provisional end-product is given a full assessment by end-users to identify modifications and improvements (Step 42). The final end-product is then created (Step 43) and released for use in the organisation using the appropriate release procedure (Step 44).

The end-product is publicised to the organisation so that all potential end-users know what it is, where it is and how it can be used (Step 45).

After the end-product has been used for some time, its impact on the organisation is assessed and documented (Step 46). Finally, a complete project review takes place to learn lessons and make suggestions that can be used to improve the methodology and support systems (Step 47).

2.5 Ensuring the End-product is Useful, Usable and Used

Let me bring this chapter to a close by describing how the different elements in the 47-step procedure contribute to three important aspects of the end-product: (i) End-users find it useful; (ii) End-users find it easy to use; (iii) End-users actually use it.

Table 2.1 shows how the steps contribute to these aspects of the end-product and the associated reasons.

Table 2.1. The impact of each step on the end-product with reasons

	Impact on the end-product			
Steps	**Useful**	**Useable**	**Used**	**Reason**
1-4	✓			Creates a good project proposal with input from all key people
5	✓	✓		Creates a k-base that is a central, structured store of knowledge
6-8	✓		✓	Defines a project scope by involving domain experts and end-users
9	✓			Identifies all sources of knowledge to be consulted during the project
10-13	✓			Creates a good project schedule that maximises the resources available
14-20	✓	✓		Uses effective methods for capturing, structuring and validating knowledge
21	✓			Reviews the project aims so they can be re-aligned for maximum benefit
22-30	✓	✓		Uses effective methods for eliciting, modelling and validating knowledge
31-33	✓	✓	✓	Creates a prototype end-product that is assessed with end-users
34	✓	✓		Reviews the project plan so tasks are focused on providing the best end-product
35	✓			Uses effective methods for eliciting deep, tacit knowledge
36-38	✓			Ensures that all of the k-base contents are correct, complete and best practice
39-43	✓	✓	✓	Assesses a provisional end-product with end-users and makes the required changes
44		✓	✓	Releases the end-product in the right way and provides information and training
45			✓	Publicises the end-product to all potential end-users
46	✓	✓	✓	Assesses the impact of the end-product and makes any necessary modifications

3

Start, Scope and Plan the Project

This chapter describes the first 13 steps in the 47-step procedure. These steps take the knowledge acquisition (KA) project from its very start, through justification and scoping to a project plan.

3.1 Step 1: Identify a Project Idea

3.1.1 Summary

In this first step, you should identify a possible project that involves a defined domain and/or named experts from which knowledge needs to be captured and stored, shared or implemented. You should ensure that the project is feasible, will benefit the organisation, and is amenable to a knowledge acquisition (KA) approach.

3.1.2 Reasons and Conditions

This step is essential since every project must start its life with an idea however vague it may be. The extent to which the initial idea is raised by the knowledge engineer or comes from other people will vary from situation to situation. It may be obvious what the project should aim to achieve or it may require work to identify this. Whatever the circumstances, it is vital that the initial idea is appropriate for the methods of KA and that a KA project can satisfy the project aims. It is a waste of people's valuable time to propose ideas that are not suitable to a KA approach.

3.1.3 Resources

The resources required for this step include: (i) Knowledge programme material, *e.g.* Knowledge business case, programme plan; (ii) General awareness material, *e.g.* of Knowledge Engineering and/or Knowledge Management; (iii) Organisation's goals, priorities and strategies; (iv) Information on previous projects and case studies; (v) Results of the organisation's knowledge audit.

3.1.4 Activities

Projects can start in all sorts of ways. Unlike later steps, it is not possible to provide a simple list of activities to follow. Instead, here are some ideas on how a good project idea can be identified.

If up and running, the organisation's knowledge programme will be educating and promoting to the appropriate parts of the organisation. The parts that are usually the most receptive to KA are staffed by people who are well-educated, have intellectually challenging jobs, are open to new ideas, are willing and able to share their knowledge and have some spare capacity. This does not mean that KA cannot be a success in other areas, but it may be less easy to accomplish.

Once aware of the benefits and limitations of KA, people will start to suggest project ideas. These can often come from 'champions' in the organisation who understand the value of acquiring knowledge and can sell the approach to others.

Poor project ideas need to be weeded out at an early stage. If the knowledge programme is operating one or more defined KA methodologies, it is important to select a project that fits the methodologies available. For instance, the organisation may be geared up to capture engineering design knowledge using MOKA but not so adept at other types of knowledge and other methodologies.

If the organisation has already conducted an audit (survey) of its knowledge assets and key experts, then this is a valuable source of information when defining and prioritising possible projects. If no audit has been conducted, then you should consider doing this as a project in its own right. Such an audit will address the following:

- Who are the key experts?
- What knowledge do they have?
- What knowledge is vital to the business?
- Which knowledge has not already been documented in an accessible, useable, understandable and useful way?
- Which knowledge is difficult to pass on?
- What hidden knowledge is there?

In running such a knowledge audit, you can use the approach described in this book, *i.e.* run it as a well-structured KA project.

If the organisation has little or no experience of KA projects, then a pilot project is a good idea. For the pilot to be a success, try to ensure that you: (i) Do something useful; (ii) Do it efficiently and professionally; (iii) Impress those involved; (iv) Show the uniqueness and benefits of a KA approach (in comparison to other approaches).

It is useful at this early stage to classify the proposed project. The categories used for this classification should be from a business perspective. For example, is it going to be the sort of project that will:

- Save money? Save time? Improve the quality of products or services?
- Solve a specific problem?
- Reduce risks? (*e.g.* a key person may leave the organisation)
- Help induction and training? Increase skills?

- Help on a specific project? (*e.g.* gather ideas for a report, gather ideas for process improvement, capture lessons learnt)

It is also useful during Step 1 to decide if the project is a domain-centred or expert-centred project. A domain-centred project focuses on a defined area of knowledge (*e.g.* let's capture knowledge of semantic web services, or hydrodynamics, or environmental legislation). An expert-centred project usually focuses on one named individual and gathers some of his/her expertise. This is often the case when someone is about to retire or is the only expert in a particular field.

By the end of this first step, you should have a strong idea for a project, which you have documented, *e.g.* as an email, as slides or in a short report.

3.1.5 Problems and Solutions

Let us look at some of the problems that can occur during Step 1 and some of the ways of preventing, eliminating or mitigating them.

3.1.5.1 Problem: Failure to Identify What the Project Will Do for the Business
To avoid this happening you must identify the specific business benefits that the project will provide. Knowledge can be a difficult thing to put a monetary value on, and the benefits of the project may be qualitative (*e.g.* risks will be reduced, service will be improved) rather than quantitative (*e.g.* 50% faster response time, 30% less material wastage). Even so, it is important to say what the qualitative improvements will be as this acts as the driver for the whole project and the benchmark against which it will be assessed at the end.

3.1.5.2 Problem: Selecting a Project that is Not Ideal for KA or for the Organisation's Methodology
To avoid this problem you must fully understand the suitability of KA for different types of projects. It is suitable for projects where knowledge already exists in people's heads or in documents. It is not suitable for projects centred purely on innovation and creativity. If the organisation has a KA methodology focused on knowledge management, then it may not be applicable to the formal, rigorous and detailed knowledge required for the development of intelligent computer systems. Equally well, if the organisation's methodology focuses on formal, rigorous and detailed knowledge, it may not be suited to the more direct, understandable knowledge required by people viewing web pages.

3.1.5.3 Problem: Selecting a Project in a Non-ideal Area, e.g. Involving Overstretched or Disinterested People
Although we as knowledge engineers try to make our projects as non-invasive as possible (*i.e.* we try to limit the disruption to the normal operations of the organisation), there is inevitably some disruption. It is vital that experts are not so overstretched that they cannot spend a few hours of spare capacity on the project. Additionally, the expert's line manager must acknowledge that a few hours per week is required from each expert. If the project will provide important benefits for the organisation, then each expert should be spared the time required for the

project. If it is not so important (*i.e.* it is a 'nice to have', background-type task), then be wary that the project will not go well.

3.1.6 Checklist for Step 1

Here are the key items that should be achieved during Step 1:

- Identify how the project can benefit the organisation;
- Identify what the project would involve and the required resources;
- Summarise the project idea in a document or presentation

3.2 Step 2: Gather Opinions on Project Idea

3.2.1 Summary

Discuss the project idea with all available key people to obtain their comments and suggestions.

3.2.2 Reasons and Conditions

The active involvement of managers, experts, end-users and support staff in the decision-making process will result in a better project idea and better buy-in from key personnel. This step should be performed on all projects.

3.2.3 Resources

The resources required for this step are: (i) Initial project outline; (ii) Business case for the knowledge programme; (iii) Awareness material for the knowledge programme.

3.2.4 Activities

The activities in this step all involve effective communication with key people. Your aim is to get people to agree to the project and support its aims. You should try to sell the project, but also listen and respond to people's concerns. You should try to involve everyone who can help the project get off the ground and become a success. You should educate, enthuse, discuss, listen, lobby, and negotiate.

You should identify the key people to see and decide on the best way to approach them. You may need a strategy to see important and busy people, such as getting buy-in from key managers whose name can be used to talk to others.

There are the usual communication channels at your disposal: phone calls, emails, face-to-face conversations, meetings and presentations.

Describe the project from a perspective that fits the person's roles, goals and responsibilities and the organisation's aims, issues, working practices and strategies. Some ideas on the issues to discuss with different roles are shown in Table 3.1.

Table 3.1. The focus of your discussion with each of the different roles

Role	Focus of Discussion
Top-level Manager	"The project will benefit the organisation in these ways…" "The project will result in these savings…" "The cost of the project will be this…"
Middle-level Manager	"The project will deliver this…" "The project will benefit your section in these ways…" "The resources required will be…" "The risks to project success are…"
Domain Expert	"Your knowledge can help other people in these ways…" "Your knowledge will be put together with the knowledge from these people…" "You will have to devote this amount of time to the project…" "The project will benefit you in these ways…"
End-user	"The project will provide you with this…" "You will be able to use it in these ways…" "It will help you when you do these activities…"
Technical Support	"The project will involve these activities…" "The project will require support in these areas…" "The time required will be this…"

When you talk to people it is important that you listen to what people ask and say. You should identify people's concerns, identify risks to project success, and identify factors to take into account for the project proposal. You may wish to use a template project proposal, which will provide points to address. Information on the contents of a project proposal is shown in Section 3.3.4 (on page 29).

A successful technique that I have used is to run a meeting and complete a project proposal form at the meeting. I use a laptop computer connected to a projector so that everyone can see the form as it is being filled-in. This helps to focus the discussions and prompts different contributions from each person. This is particularly helpful in generating a consensus view and in gaining agreement from all parties. I would normally allow 2 hours for such a meeting.

3.2.5 Problems and Solutions

3.2.5.1 Problem: Not Involving Experts and Their Line Managers at This Stage
It is tempting for a knowledge engineer not to involve the domain experts at such an early stage in the project's life. It is even more tempting not to involve each expert's line manager. The reasons why these things may happen can include the knowledge engineer: (i) Feeling embarrassed at his/her lack of knowledge in the expert's domain; (ii) Not wishing to waste the valuable time of experts and managers; (iii) Not knowing who the experts and managers are; (iv) Lacking

confidence in approaching people who may be more senior; (v) Not realising the importance of involving experts and managers at this stage; (vi) Fearing that the project may be unduly criticised before it has had time to mature. These are all valid reasons, but attempts must be made to overcome them if the project is to be a success. Remember that the project will be made much easier and worthwhile later if you involve the people who know most about the domain.

3.2.5.2 Problem: Getting Conflicting Comments from Different People

It is almost inevitable that different people will have their own ideas and these may be conflicting. The experts know a great deal but may not realise how much they know. Their ideas on what knowledge to capture and how it can help the organisation are valuable but may not fit with what managers think, since the latter will have more of the bigger picture. This, in turn, may be different again from the concerns and needs of the end-users. The important thing is to gather as many opinions as possible. Look for common ground. Treat everyone's input as genuine. Realise that people have their own views on the project. When two views directly conflict and you cannot decide which to select, then consider including both as possible options to be explored later.

3.2.5.3 Problem: Not Listening and Responding to People's Concerns and Ideas

As already discussed, everyone's input should be treated as genuine and useful input. You are starting to make major decisions that will affect the direction in which the project will move. Decision-making is often not straightforward and can sometimes come down to gut feel and even guesswork. That said, you must be able to defend the decisions you make with good reasons. It may be the case that you have to side more with one person's ideas and downplay the input from others. If this is the case, then make sure you have good reasons for the decisions you took.

3.2.6 Checklist for Step 2

- Explain the project idea to all relevant people;
- Gather comments and suggestions from all relevant people;
- Identify factors required for the project proposal

3.3 Step 3: Create a Project Proposal

3.3.1 Summary

Create a document that describes the reasons for the knowledge acquisition project, what it aims to achieve, the resources required, and other key issues. You might refer to this as a project proposal, a business case, or another relevant term.

3.3.2 Reasons and Conditions

The business case and resource requirements for the project need to be clearly documented to ensure all key personnel understand the project aims. This enables them to make an informed assessment of the project so as to make the appropriate decisions. The project proposal acts as a contract stating what the project intends to do for the organisation within the defined resources. It is also a benchmark against which to assess the project deliverable when it is being used.

A project proposal should be created for all projects. The only possible exceptions are for projects that are very straightforward, off-the-shelf ones that have been performed many times in the organisation. Examples of this could be a retiree project or a new starter project.

3.3.3 Resources

The resources required for this step are; (i) Information gathered in Step 2; (ii) Proposal template; (iii) Example proposals.

3.3.4 Activities

There is a single activity for this step, *i.e.* write the project proposal. To do this, you should make use of the information you obtained during Step 2 together with any proposal templates or standards. It is usual for the proposal to use a standard format such as sections on a form or headings on the rows of a table. The headings that are often used are shown below. Also shown are some questions that need to be addressed for each section.

- **Objectives:** What will the project do for the organisation? What problems will it address? What improvements will it make? What savings will it result in?
- **Deliverables:** What will the project produce? (A website? A document? An ontology?). Broadly speaking, what will the deliverable contain?
- **Project Personnel:** Who will be the knowledge engineer/s? Who will be the project manager? Are other people required to support the project or the deliverable?
- **Domain Experts:** Which people possess the knowledge that needs to be captured?
- **End-users:** Who will use the deliverable?
- **Other Resources:** What other resources are required? Equipment? Meeting rooms? Travel? Technology?
- **Business Metrics:** What can be measured to assess the success of the project when the deliverables are deployed?
- **Funding:** What are the estimated project costs? Is funding in place to cover the project costs?
- **Timescales:** When will the project finish? On what dates will key milestones be complete?
- **Risk Analysis:** What are the main project risks? If a problem occurs, what will be the impact? What can be done to avoid problems? What can be done to deal with a problem if it does occur?

Using this format, an example of a project proposal is shown in Figure 3.1.

Project Proposal: Semantic Webs Services

Objectives: This project will acquire and share knowledge concerning semantic web services. These new web technologies are likely to have a major impact on our business; hence we need to understand their potential for improving our procedures and technological infrastructure. A main objective is to enable managers to make better decisions in this area.

Deliverables: The project will create a knowledge web describing the semantic web and semantic web services. It is likely to focus on both the technological side and the business applications of this new technology.

Project Personnel: The knowledge engineer on the project will be James Taylor. The project manager will be Sharon Swift. The web administrator will be Paul Jacobs. IT secretaries will transcribe interview recordings.

Domain Experts: The main experts reside in three locations: (i) University researchers working on semantic web services, (ii) Consultants currently working with us on IT technologies, (iii) In-house experts (within the IT Group).

End-users: Potential end-users are: (i) All managers associated with engineering, technology and IT, (ii) The IT Group, (iii) Engineers and technicians, (iv) The Knowledge, Information and Logistics Group.

Other Resources: There will be a need to interview experts in several universities around the country. Hence, a web-enabled meeting technology would be beneficial for some of the acquisition and validation sessions.

Business Metrics: There are no direct, quantitative measures of success for a project of this kind. Instead, assessment will use a range of qualitative and indirect measures resulting from end-user questionnaires, intranet statistics and feedback sessions.

Funding: The main project costs are: (i) 8 weeks of effort from the knowledge engineer; (ii) 1 week from the project manager; (iii) 2 weeks from domain experts. It has been agreed that this will be funded under the existing IT development budget for the current year.

Timescales: The project will take an elapsed time of 3 months. Assuming a start date of mid-July, it will be complete by mid-Oct. Key milestones are: Project Plan (end July), Completion of knowledge base (end Sept).

Risk Analysis: Major risks to the project are: (i) Unavailability of some experts due to summer vacations. This will be avoided by pre-planning dates and using experts with overlapping areas of expertise; (ii) Lack of availability of knowledge engineer. This can be avoided by the continued commitment from the IT Director for this project and its priority with respect to other tasks; (iii) Lack of use of website when complete. This can be avoided by involving end-users in the design and development of the website and providing training and awareness sessions when necessary.

Figure 3.1. Example of a project proposal

3.3.5 Problems and Solutions

3.3.5.1 Problem: Proposal is Unclear

It is vital that the proposal lays out the case for the project in a clear and specific way. If you are struggling to do this, then you may not have a good project idea and may need to think again and revisit Steps 1 and 2. If you are convinced that you have a strong project idea but are having difficulties creating the proposal then try one or more of the following: (i) Look at previous project proposals; (ii) Use any standard proposals or templates that are available; (iii) Ask someone from the knowledge support team for their help and advice; (iv) Rather that trying to write something in clever language or business-speak, just use plain language to convey the necessary information; (v) Use lists or bullet points rather than sentences.

3.3.5.2 Problem: Proposal Misses Out Key Information

It is important that the proposal has all the necessary information about the project. Look at previous project proposals and any available templates to check you have included all the relevant points. If there is an experienced knowledge engineer available in the support team, then ask for his/her assistance. The knowledge programme may have set-up a formal review at this stage that will help assess and improve the proposal.

3.3.5.3 Problem: Proposal Oversells the Project and Exaggerates the Objectives and Deliverables

It is all too easy to get carried away with a project and over-exaggerate what it will do for the organisation. Try to put the project in context and provide realistically attainable goals. It is better to slightly undersell the project so that people will be pleasantly surprised when you deliver more than they were expecting.

3.3.6 Checklist for Step 3

- Document the project objectives, deliverables and metrics;
- Document the project personnel, their roles and responsibilities;
- Document the time and cost of the project, and project risks

3.4 Step 4: Get Agreement for Proposal

3.4.1 Summary

Present or circulate the project proposal to key people, make modifications where necessary and obtain agreement, commitment, resources and go-ahead for the project.

3.4.2 Reasons and Conditions

All key people must agree to the project aims and resource allocation. If not, serious problems can occur later due to a lack of resources, priority and understanding of the project. This step must be done on all projects.

3.4.3 Resources

The main resource for this step is obviously the proposal you wrote in the last step. The other resources are people and channels of communication.

3.4.4 Activities

The first activity is to identify the key people to involve in this step. As a bare minimum, the designated project manager and one of the key domain experts should be involved. The personnel who are likely to form the project team should also be involved. This will include domain experts, end-users and members of the knowledge support team. To get resources allocated, it is essential to involve the line/resource managers of the project team members, *e.g.* managers of the experts and end-users.

The next task is to circulate the proposal and gain feedback. The method of communication can often be determined by the size of the project.

- For a short project (less than 3 months), it is acceptable to communicate with people on an individual basis. This can be done with face-to-face get-togethers, by email or by phone;
- For larger projects (lasting more than 3 months), a meeting is probably the best way to present and discuss the proposal, gain commitment and establish funding;
- For the average project of around 3 months, the choice of method will depend on the circumstances and your particular preferences

Whatever the method, it is vital that each person (especially the project manager) is fully informed, fully committed and has agreed to resource the project. Of course, at this stage, the exact amount of time required from the project team (especially the experts) is unknown, since the exact areas of knowledge to be acquired have not be defined in detail. However, broad outlines of people's commitment can usually be made.

If any changes to the proposal are made during this step, you should circulate a finalised proposal to all key people and any others who may have an interest in supporting or benefiting from the project.

3.4.5 Problems and Solutions

3.4.5.1 Problem: Not Enough People Read and Comment on the Proposal
You cannot force people to read your proposal or to attend a meeting to discuss it. Some people may be too busy or disinterested, particularly if it is a short project that is not seen as vital to the organisation. Try emailing the proposal followed by a short phone call. In this way, you can talk the person through the proposal and gain their comments without taking too much of their time. If required, get the project manager to talk to the person's line manager. However, be warned, if key people are too busy to be seen or too disinterested in the project at this stage then perhaps the project should be delayed, modified or abandoned.

3.4.5.2 Problem: People Seem Interested but Do Not Give Full Commitment
or Do Not Allocate Adequate Resources
People in organisations are often overloaded with work. Your project has to
compete for people's time with many other projects and activities. Be realistic.
Identify what resources are available for your project and try to work within this. If
necessary, talk to the project manager about making the project goals less
ambitious. Bear in mind that however much a person says he/she is too busy to be
involved, most people can spare one or two hours per week. This is generally the
maximum that is required from each person on the project team (other than the
knowledge engineer).

3.4.5.3 Problem: One or More Key People Fail to Give Commitment
to the Project, but the Project Proceeds Anyway
In an ideal world, everyone would understand what your project is trying to
achieve, be able to accurately assess if it is possible, and make the necessary
decisions. Life is rarely like this, and for some projects it may not matter too much.
If you have enough commitment from enough people, then continue with the
project, but keep hammering away at the people who have not provided full
commitment. Send progress reports to these people, and demonstrate prototype
deliverables when available.

3.4.6 Checklist for Step 4

- Identify the key people and circulate the project proposal to them;
- Listen to comments and feedback and make modifications where necessary
  to achieve an agreed project proposal;
- Ensure that key people are fully committed to the project and have
  allocated the resources required to fulfil the project aims

3.5 Step 5: Create a K-base

3.5.1 Summary

Create a new knowledge base (k-base) or define a new k-base partition (if your
organisation is using a single, enterprise-wide k-base). The k-base is the store into
which the knowledge is populated during the project. The k-base is a special type
of database and is usually coded using modern web formats.

3.5.2 Reasons and Conditions

It is useful to have a k-base at this early stage of a project so you are able to start
populating it with useful knowledge during the next few steps. If you are using a
sophisticated tool to create your k-base, especially one that has a workflow tool to
lead you through a project, then you could do this step earlier in the project (*e.g.* to
help create the project proposal).

3.5.3 Resources

The resources required for this step are: (i) Tool to create the k-base; (ii) K-base template, *i.e.* a generic ontology.

3.5.4 Activities

The activities here are dependent on the technology available in your organisation for storing knowledge. If you have a dedicated KA tool then things are made significantly easier. If not, then you must identify what is available. If you have the choice then you should be looking for a tool that can:

- Store knowledge in an object-oriented way so that each item of knowledge (*i.e.* concept) can have relationships to other items and can have properties (attributes and values);
- Make use of modern web formats such as XML, RDF and OWL (Fensel, 2004) that allow the k-base to be published, transformed and used in a variety of ways;
- Create a k-base using a generic ontology that will provide a standard structure for the k-base. The generic ontology should define high-level classes of concepts, a set of relations and a range of formats for representing the knowledge;
- Allow the k-base to be viewed in different ways using template-driven k-models. This will vastly improve your efficiency at capturing, validating and sharing the knowledge

You might consider using a normal database, or even a word-processor or spreadsheet file for the k-base. However, these options should only be used for small k-bases when there is no funding in place to purchase dedicated tools. Such a primitive k-base (if it can be called a k-base) is unlikely to allow multiple ways of viewing the knowledge, which is useful for editing and presenting the knowledge (*e.g.* for validation with experts). Primitive k-bases will also necessitate the use of a separate website design system if the knowledge is to be made available on the organisation's intranet.

If your organisation has a single, enterprise k-base then you will have to define an area (partition, segment) of the k-base for you to populate and edit. The knowledge support team in your organisation should have defined the way in which this is achieved.

An important activity is to select the generic ontology to use, *i.e.* a skeletal structure on which to start the k-base. An example of a generic ontology called GTO is described in Appendix 1. GTO, like other generic ontologies, gives you a standardised way of starting a k-base that will help in a number of ways:

- Like any template, you start with something to fill-in, not just a blank canvas;
- Standardising k-bases makes it easier:
  - To see what other people have done;
  - For the support team to assess the k-base and provide assistance;
  - To create a best practice approach

- It makes knowledge capture and modelling more efficient by:
    - Reducing the time taken to start populating a k-base;
    - Making it easier for the knowledge engineer to use the KA tool;
    - Prompting for knowledge gaps that might be acquired

- It can help reduce the time and difficulty in training new knowledge engineers;
- It can standardise the structure and navigation of knowledge webs;
- It allows an enterprise k-base to have a consistent structure

Note that a generic ontology is often a starting point and not the final structure that the k-base will end up having. The exact structure of the k-base will emerge and develop as you perform the interviews, analysis and modelling in Steps 14 to 24.

3.5.5 Problems and Solutions

3.5.5.1 Problem: Technology Used for the K-base is Inadequate for the Project
Unfortunately, those responsible for purchasing the technological infrastructure and support tools required for KA do not always have the necessary understanding to obtain the right technology to help the knowledge engineer. Nor do they always have the funding in place to purchase specialist tools. There is a huge advantage to be gained by using dedicated tools and the right k-base technology in terms of the efficiency of the knowledge project and effectiveness of its deliverables. If you are using inadequate technology, then you should still try to obey the right principles of good KA, such as the use of k-models, creating a re-usable k-base, storing all the knowledge in a single place (file, directory) and ensuring there are no inconsistencies (*i.e.* contradictory knowledge) within the k-base. With a dedicated tool, such as PCPACK, you do not have to think too much about these things as the tool is designed to ensure good practice.

3.5.5.2 Problem: Unavailability of IT Support/Administrator to Create K-base
Organisations can vary in the privileges they allow its knowledge engineers to create or define new k-bases. If you need someone else to do this, then you rely on his/her availability and motivation to do what you need. If you have to wait, then it is better to get on with the next few steps rather than hold up the project. You can save the knowledge captured in a word-processor file until the k-base is ready for you to use.

3.5.5.3 Problem: Unavailability of Technological Infrastructure
You may have the right technology and tools on-site but you may not have access to it, *e.g.* you are a new knowledge engineer, or the software is still being released or the number of licenses limits the number of users. It is important, therefore, that you see the people that matter to get yourself access to the technology as quickly as possible. If it cannot be accessed from or loaded onto your desktop PC, then find out why and get your line manager to push the IT support staff to do the necessaries. In the meantime, you may have to beg, borrow or loan out a laptop or another PC to use. Alternatively, do what you can while you are waiting by using existing tools and populate the k-base when it becomes available to you.

3.5.6 Checklist for Step 5

- Identify the technology to use for the k-base (if not already in place) and the k-base partition if using an enterprise-wide k-base;
- Decide on a structure (template, generic ontology) for the k-base;
- Create or request creation of the k-base or k-base partition, and check that it is working and that access for editing is unhindered

3.6 Step 6: Decompose Domain

3.6.1 Summary

Decompose the domain into topics (*i.e.* key knowledge areas) by interviewing one or more experts. This is the first of two steps that taken together provide a project scope, *i.e.* the specific parts of the domain knowledge that will be acquired during the project.

3.6.2 Reasons and Conditions

No project can capture all of the knowledge to a deep level of detail. Given the project resources and aims, you need to identify the essential knowledge to capture, store and share. To do this you must first interact with one or more experts to break the domain into different topics so that each topic can then be rated and assessed for selection.

3.6.3 Resources

The main resources required for this step are one or more experts. Other resources are: (i) Information contained in the project proposal; (ii) Information contained in the organisation's knowledge audit or capability database (if these exist).

3.6.4 Activities

The first activity is to arrange one or more scoping sessions with the experts. If you hold a single session, then it could cover this and the next step. To arrange the meeting, you need to decide how many experts to involve during the scoping session/s. There are no right or wrong answers when deciding this. If there is only one main expert in the domain, then the choice is made. If there are multiple experts with differing areas of expertise, differing experiences or differing roles, then it may be useful to have a group session with 2, 3 or 4 experts. An advantage of this is that more experts will feel involved in deciding the direction of the project. On the downside, the session is likely to take longer since experts can disagree on the best way to view and decompose the domain. Involving two experts in such a case is often a good compromise.

During the session, you should do something like this:

- Explain to the expert/s about the need for scoping, *i.e.* that you need to focus the project on the areas of knowledge that will maximise the benefit of the project to the organisation, and to do this you first need to break the domain into topics, then rank or rate each one;

- Remind the experts of the project aims and other issues on the project proposal;
- Talk about different types of topics:
    - Knowledge Areas (*e.g.* subjects, skills);
    - Main tasks (performed by the experts or by the organisation);
    - Projects, Products, Technologies; Major issues.

- Use a tree diagram, or a structured-listing method, to display some topics and, if necessary, some sub-topics of these;
- Review and discuss the topics until you have identified between 5 and 20;
- Assess the amount of knowledge within each topic as compared to the other ones;
- If required, modify the topics so that none of them are a lot larger nor a lot smaller than the others (by breaking larger topics in half, and aggregating smaller ones together);
- Ask the expert/s to explain in simple terms what each of the topics means if this is not already known or is not obvious (and take an audio recording as the explanation may contain useful information to put in the k-base);
- Using input from the expert/s, try to assess the size of the topics compared to the project resources: How many can be acquired during the project? To what level of detail? Some data to help you do this is shown in Appendix 2. This assessment will be re-visited in the next step (Step 7).

As an illustration of the different types of topic that you might identify, three are listed in Table 3.2. The domain here is knowledge acquisition. Note that for an actual project in the domain of knowledge acquisition, only one of these topic sets would be chosen for the next stage.

Table 3.2. Three possible topic sets for the domain of knowledge acquisition

Knowledge Areas	Tasks	Elicitation Techniques
Knowledge Models	Initiate Project	Unstructured Interviews
Knowledge Objects	Scope Project	Semi-Structured Interviews
Elicitation techniques	Plan Project	Laddering
CommonKADS	Plan an Interview	Concept Mapping
MOKA	Perform an Interview	Process Mapping
GTO	Analyse the Knowledge	Teach Back
Web Technologies	Model the Knowledge	Repertory Grid
Knowledge Management	Validate the Knowledge	Concept Sorting
Psychology of Expertise	Elicit Tacit Knowledge	Triadic Elicitation
Knowledge Stores	Publish Knowledgebase	Commentary

3.6.5 Problems and Solutions

3.6.5.1 Problem: Domain is Decomposed in the Wrong Way
This can be prevented in a number of ways. First, ensure you talk to the experts and get their feel for the way in which the domain can be split. This is sometimes not easy, especially for experts who do not think in a structured way. Second, try different ways of decomposing the domain and see which one seems the best. Third, use the information on the project proposal to drive how you split up the domain, *i.e.* base it firmly on the objectives of the project. For example, if the project is very task-oriented (capturing knowledge of how the experts perform activities) then use tasks for the topics. Fourth, use generic ontologies to help prompt the expert for different topic areas. For example, you might use GTO (explained in Appendix 1) as a way of suggesting ideas and capturing some initial knowledge.

3.6.5.2 Problem: The Number of Topics is Too Few or Too Many
Having only 3 or 4 topics is usually too few to provide enough granularity to allow a good choice to be made. If the experts are involved in 3 or 4 main tasks, or know about 3 or 4 main areas, then try sub-dividing each into 2 or more sub-tasks or sub-areas, so as to increase the number of topics. If you have too many, then the task of rating and selecting will take too long. There are two ways to reduce the number of topics. First, dispense with some that are obviously not going to benefit the organisation (*e.g.* knowledge that is obsolescent, infrequently used or already documented in detail). Second, try grouping topics together to form larger topics.

3.6.5.3 Problem: Topics are of Different Sizes or Types
Differing sizes is not a major problem, which is just as well as it can be difficult for the experts to ascertain how much knowledge is involved in a particular topic (which is usually only revealed when the elicitation interviews take place). The important thing at this stage is to encourage the expert to estimate the sizes of the topics in relation to one another. If the topics are of different types, then this need not be a problem as long as the topics taken as a whole represent the total knowledge that could be acquired (if resources were limitless). If not, then you may have topics that have a lot of overlapping knowledge or taken as a whole have gaps. If so, you should encourage the expert to decompose the domain in another way, or to add new topics to fill in the gaps.

3.6.6 Checklist for Step 6

- Select a way of breaking down the domain knowledge (*e.g.* by tasks, by skills, by technologies) into a number of topics;
- Create a tree containing between 5 and 20 topics (*i.e.* sub-domains of knowledge);
- Ensure that the topics are roughly the same size and assess the size relative to the project resources

3.7 Step 7: Rate and Select Knowledge Areas

3.7.1 Summary

Rate the candidate areas of knowledge (topics) using input from key people and then select those areas that will be acquired during the project.

3.7.2 Reasons and Conditions

This is the second step in the scoping procedure. It is required so that the resources allocated to the project are focused on acquiring knowledge that will result in maximum benefits. Input from experts is essential to rate and select the most important areas of knowledge. This step should be performed on all projects.

3.7.3 Resources

The resources required for this step are: (i) The topics defined in Step 6; (ii) Key members of the project team, *i.e.* one or more experts, one or more end-users, and the project manager; (iii) Information on the organisation's strategies, plans and priorities.

3.7.4 Activities

There are four main activities: (i) Identify the people to be involved and arrange one or more sessions with them; (ii) Rate or rank the topics in terms of benefits to the organisation; (iii) Rate or rank the topics in terms of knowledge acquisition; (iv) Use the ratings/rankings to identify the scope of the knowledge to be acquired during the project. Let us look at each of these.

3.7.4.1 Identify the People and Arrange the Session/s
In the first main activity you should decide who to involve in rating/ranking the topics and who to involve in identifying the scope. Domain experts are the major source of specific knowledge on each topic, hence are often involved. End-users can also be considered, as they will be the ones using the knowledge in the end-product. The project manager is another possibility, especially if he/she has detailed domain knowledge. The more people you involve, the more of a rounded picture you will gain. However, if you intend to run a single session, then a lot of attendees can provide a lot of different opinions. If you do intend to involve more than about 3 people, then I suggest seeing people individually and collating their individual views afterwards.

3.7.4.2 Rating or Ranking in Terms of Organisational Benefits
Once you have identified who to involve, the second main activity is ask these people to rate or rank each topic in terms of the benefits to the organisation of acquiring and using the knowledge in the end-product. This can be done informally by ranking the topics using one or two key dimensions. Hence, the ranking would sort the topics in terms of one or two key questions, such as: Which topics would provide the most help to the end-users? Which topics would save the most money for the organisation? Which topics would reduce the time taken to perform

essential activities? Which topics would reduce risks? Which topics would enable people to increase their skills? Obviously the dimension or dimensions on which to sort the topics would be derived from the project proposal.

One way of doing this is to write each topic on a card and ask the people involved to simply sort the cards into a line so that one end represents the most useful topic and the other end represents the least useful topic. Another way is to ask the people to sort them into a number of piles, *e.g.* a 'very important' pile, a 'medium important' pile and a 'less important' pile.

A more complex, but more thorough, method is to use several different dimensions and ask people to provide ratings (*i.e.* scores). These ratings can then be used to sort the topics using weighted averages. The easiest way for me to explain how to do this is to describe an example. This is illustrated in Figure 3.2.

Tasks	Cost savings	Time savings	Less risk	Increased quality	Weighted Average	Rating of Importance
Weighting →	*3*	*1*	*2*	*2*		
Plan route	1	2	1	2	2.75	Less Imp't
Book flights	3	3	2	1	4.5	Very Imp't
Decide destinations	2	2	2	2	4.0	Medium Imp't
Book hotels	3	2	2	3	5.25	Very Imp't
Arrange insurance	2	1	2	1	3.25	Less Imp't
Arrange visas	1	3	3	1	3.5	Medium Imp't
Arrange transit	2	3	3	2	4.75	Very Imp't
Arrange trips	2	2	1	3	4.0	Medium Imp't

Figure 3.2. Grid showing importance to the end-user of each topic

The following activities would generate the kind of table shown in Figure 3.2:

- Write the topics in the first column. In the example in Figure 3.2, they are 8 tasks involved in arranging a vacation abroad.
- Identify a set of dimensions that provide benefits and write them as headings for columns. In the example in Figure 3.2, there are 4 dimensions: Cost savings, Time savings, Less risk and Increased quality.
- Ask the people involved to provide weightings for the dimensions, *i.e.* the quantify how much benefit would derive from each dimension. In the example in Figure 3.2, the dimension 'Cost savings' has been given the weighting 3 showing it is a very important aspect of the project's end-product, whereas 'Time savings' has been given the weighting 1 showing it is a less important aspect.
- Ask the people involved to provide ratings of each topic against each dimension (ignoring the weightings). A rating of 3 represents a high

benefit, a rating of 1 a low benefit. For example, in Figure 3.2 the topic 'Plan route' has the rating 1 for 'Cost savings' and the topic 'Book flights' has the rating 3 for 'Cost savings'. This means that capturing knowledge of planning routes and using it in the end-product would provide little savings in costs for the end-users (or the organisation), but capturing knowledge of booking flights would provide good costs savings.

- Once all the ratings have been made, they can be combined with the weighting to provide a weighted average. For example, in Figure 3.2, the topic 'Book flights' is given the weighted average of 4.5 because this is the result of the sum (3x3)+(1x3)+(2x2)+(2x1) divided by 4.
- To complete the table, the weighted averages are sorted into three groups so that the highest third are rated as 'Very Important', the middle third as 'Medium Important' and the bottom third as 'Less Important'. This is shown in the final column of the table.

3.7.4.3 Rating or Ranking in Terms of Knowledge Acquisition

The third main activity is to rate or rank each of the topics in terms of the ease of acquiring and implementing the knowledge. This would be done in the same way as you did the rating/ranking against organisational benefits except that the dimensions are different. Here the dimensions represent questions, such as: Which topics would be the easiest to elicit from the experts? Which topics would involve the most available experts? Which topics would involve the least tacit knowledge?

Just as before this can be achieved by sorting cards into a line, into piles or using the more complex procedure of ratings and weighted averages. Figure 3.3 shows an example of a grid of ratings using the more complex procedure in the same domain as used in Figure 3.2.

Tasks	Availability of experts	Explicit knowledge	Existing docs	Generic knowledge	Weighted Average	Rating of easiness
Weighting →	3	2	1	2		
Plan route	2	1	1	3	3.75	Medium
Book flights	2	3	2	2	4.5	Easy
Decide destinations	1	2	3	2	3.5	Hard
Book hotels	1	2	2	1	2.75	Hard
Arrange insurance	3	1	2	1	3.75	Medium
Arrange visas	1	2	1	2	3.0	Hard
Arrange transit	2	2	2	2	4.0	Medium
Arrange trips	2	3	1	2	4.25	Easy

Figure 3.3. Grid showing the ease of capturing and implementing each topic

3.7.4.4 Identifying the Scope of the Knowledge to Be Acquired

The final main activity in this step is to identify which topics will be acquired during the project for use in the end-product.

To do this, you should first collate the information produced during the rating and ranking sessions. If you used a single dimension to sort the topics, then this can be a simple matter of calculating average values from each of the people that were involved (assuming you saw them individually).

If multiple dimensions were used, then a weighted average can be used by assigning weights to the different dimensions and calculating an average (as was explained earlier).

If you have used the complex method that created two tables (as illustrated in Figures 3.2 and 3.3), or another method using 2 dimensions, then you can present the results using a 3x3 matrix. An example of this is shown in Figure 3.4, which uses the data contained in Figures 3.2 and 3.3 to display the results. As shown, in Figure 3.4 the two factors used during the scoping form the rows and columns of the matrix. Each cell of the matrix is populated by the topics that were ranked or rated on the two factors.

		Ease of developing the help system		
		Easy	**Medium**	**Hard**
Importance for end-user	**Very Important**	Book flights	Arrange transit	Book hotels
	Medium Importance	Arrange trips		Decide destinations Arrange visas
	Less Important		Plan route Arrange insurance	

Figure 3.4. Example of a scoping matrix

Once you have presented the results of the rating/ranking activities, you should decide how many topics to select. This will depend on three main factors:

- The amount of time available for the project;
- The size of each topic (*e.g.* as a percentage of an expert's overall span of knowledge);
- The depth to which the knowledge needs to be acquired (*e.g.* is a basic level required, or a very detailed understanding?)

For example, if each topic represents a substantial area of knowledge and you need to capture it in great detail (*e.g.* for coding into an intelligent system), then you may only be able to acquire one or two topics in a typical 3-month project.

Information on the amount of time to acquire different amounts of knowledge is shown in Appendix 2.

Using the information in Appendix 2 and other information at your disposal, you can determine the number of topics to select. Using this and the rating/ranking data will allow you to select the topics that will be your proposed scope.

If you are using a 3x3 matrix, as shown in Figure 3.4, then it is usually straightforward to decide on a proposed scope for the project. Any topics in the top-left cell should be in the scope and any in the bottom-right should be outside the scope. Topics in the top-middle cell (medium easiness and very important) and middle-left cell (easy to do and medium importance) would usually also be included in the proposed scope.

To illustrate this, using the example shown in Figure 3.4, let us assume that there are three topics that can be selected from the eight in this example. In this case, *Book flights*, *Arrange transit* and *Arrange trips* would be in the proposed scope and the others would not. Those in the right-hand column would be excluded as being too hard and taking too many resources. Those in the bottom row would be excluded as being of not enough importance for the end-user.

3.7.5 Problems and Solutions

3.7.5.1 Problem: Rating Takes Too Long
Rating (scoring) can be a lengthy procedure since the number of ratings to make is equal to the number of topics multiplied the number of rating factors. There are a number of ways of tailoring the procedure that can shorten the procedure. First, use a ranking method rather than ratings. Here, ask the attendees at the scoping meeting to place the topics in a priority order rather than provide scores for all of them. Second, only rate a sub-set of the total set of topics. This means pruning a number of topics before rating begins. This can be done by asking the attendees to discard the least important or use a ranking method. Whichever method is used, it is vital to ensure that scoping is done thoroughly as it is one of the most important parts of a KA project.

3.7.5.2 Problem: Attendees Say Everything is of the Same Importance and/or Have Difficulty Assessing the Importance of the Knowledge
If the attendees find it difficult to assess the differential importance of the topics, then rating or ranking methods are good ways of helping them to separate the topics into a prioritised order.

3.7.5.3 Problem: The Most Important Areas of Knowledge to Focus On are Also Those that Will Be the Most Difficult to Acquire, Model and Share
It is rare that every single topic that is rated as 'very important' is also rated as 'difficult to acquire'. However, this may happen and if it does, then difficult decisions will have to be made. For instance, it may be better to avoid any topic that is 'difficult to acquire' (even if it is 'very important') and select topics that are 'important' and 'easy to acquire'. Alternatively, select just one or two 'very important' topics even if they are 'difficult to acquire'.

3.7.6 Checklist for Step 7

- Rank or rate each topic against one or more dimensions (such as benefits to the organisation and ease of acquisition);
- Decide on the number of topics to be included in the scope;
- Select a set of topics that will form the proposed scope

3.8 Step 8: Agree Scope and Finalise Aims

3.8.1 Summary

Present the scope identified on the previous step to key decision-makers and (if required) make changes to finalise the project focus, aims and deliverables.

3.8.2 Reasons and Conditions

The proposed scope defined in Steps 6 and 7 needs to be presented to key people so they can provide comments in light of the issues and goals in the organisation. It is vital that key project personnel, especially the project manager and representatives of the end-users, are happy with the scope.

3.8.3 Resources

The resources required for this step are: (i) The project team, *i.e.* project manager, experts, end-users, managers; (ii) The proposed scope in the form of a short document or presentation.

3.8.4 Activities

The first task is to decide whether to contact people individually or at a group meeting. If the project is relatively small (less than 3 months), then you can present the scope to people on an individual basis. For larger projects (lasting more than 3 months), a scoping meeting is advisable so that the proposed scope can be presented and discussed. For the average project of around 3 months, the choice of method will depend on the circumstances and your particular preference.

If you are presenting to people on an individual basis, then you should write and circulate a scoping document, then follow this with individual conversations. If you are running a scoping meeting, then you should create a presentation. In both cases, the information to be presented should:

- Describe the basic procedure you used, *i.e.* who did you see and what did you do;
- Describe the topics that were identified;
- Present the ratings or rankings that were assigned to each topic;
- Present the results and the proposed scope

A matrix is a clear way of presenting the results and the proposed scope. An example of this was shown in Figure 3.4.

The next activity is to circulate the scoping document and/or run the scoping meeting. The agenda for a scoping meeting would be something like this:

1. Introduction (Describe what scoping is and why it is needed);
2. Scoping method (What was done to arrive at the proposed scope);
3. Results (*e.g.* Presented as a 3x3 matrix);
4. The proposed scope (*e.g.* Upper-left portion of the 3x3 matrix)
5. Discussion;
6. Document the finalised scope and any outstanding actions

The scoping meeting should end with a decision upon the scope, which should be documented for all to see and agree to. It can be useful to fill-in a computer-based form, which is projected so that all can see it take shape. Here are the sorts of items to have on the form:

- **Knowledge to be Captured:** What specific areas of knowledge will be captured during the project?
- **Sources of Knowledge:** Who are the domain experts? Which documents, or other sources of knowledge, are relevant and available? (See Step 9)
- **Expert-Domain Map:** Which experts have expertise in the different areas of knowledge in the scope.
- **Aims and Format of the Deliverables:** What are the specific aims of the deliverables? What will the deliverable contain? Who will use the deliverable and how will they use it?
- **Risks and Mitigating Actions:** What are the specific project risks? If a risk occurs, what will be the impact? What mitigating activities can take place?
- **Operational Metrics:** What specific measures can be used to assess the benefits and impact of the end-product?

After the meeting, this form should be circulated to other relevant people who were not in attendance.

If you have communicated with people individually, rather than at a meeting, then such a form should be filled-in by yourself and then circulated. This will allow all key people to be aware of the scope.

3.8.5 Problems and Solutions

3.8.5.1 Problem: Key Decision-makers Cannot Agree on a Suitable Scope
Reaching consensus and creating compromise are elements of all projects. Holding a meeting between key individuals is often a good way of doing this. At the meeting, each person's views should be understood and analysed in the light of the project aims and circumstances. If consensus cannot be agreed then the project manager should have the final say on the project scope. You, the knowledge engineer, should be in agreement and feel confident that: (i) It will be possible to capture the knowledge within the timescales; and (ii) The captured knowledge will be useful to the organisation.

3.8.5.2 Problem: The Proposed Scope Identified in Step 7 is Taken as Read or Ignored and Replaced by a Large and/or Fuzzy Project Aim
It is always possible that the scope developed by talking to experts is not going to be exactly the one required by the organisation as a whole. Rather than rejecting any other ideas, they should be considered. The proposed scope defined in Step 7 is a proposal and should be open to tweaks and changes. It certainly should not be replaced by something non-specific or too large for the project resources. The proposed scope should be treated as a good, first-pass definition of the exact areas of knowledge to acquire during the project. On that basis, Step 8 is concerned with gaining agreement and making any necessary modifications.

3.8.5.3 Problem: Cannot Get Time With, Or a Response From, Key Decision-makers

If people are unavailable or too busy then you cannot force them to be involved in the scoping process. This may be a sign that the project is not right for the organisation. If so, consider delaying or abandoning the project. If you decide to go ahead, communicate with the people who are interested, but do not ignore the people who are unwilling to attend meetings or send replies to emails. Keep them up-to-date on the scoping decisions using emails or other channels. If they complain later that the project is not doing what it should, then you have the grounds to respond (*i.e.* they have been kept fully informed about the scope and did not provide input when they had the opportunity).

3.8.6 Checklist for Step 8

- Send or present the proposed scope to key decision-makers and gather comments;
- Agree a scope with all key decision-makers;
- Finalise and document the scope, *i.e.* the areas of knowledge to acquire

3.9 Step 9: Identify Sources of Knowledge

3.9.1 Summary

Identify where you will get the knowledge. This can include one or more of these: experts, end-users, documents, databases, websites and IT systems.

3.9.2 Reasons and Conditions

You need to identify the people and places where the knowledge resides so that you can properly plan the project and avoid wasting effort in gathering the knowledge. For each particular topic, it is very useful that you understand two things: (i) Where it will be gathered from; and (ii) Which expert will validate (check) the knowledge after it has been captured. This step should be performed on all projects and can often be combined with the previous step, *i.e.* the information is gathered at the scoping meeting.

3.9.3 Resources

The resources required for this step are: (i) Project Scope; (ii) Project objectives; (iii) Experts; (iv) Knowledge resources such as documents, databases and websites.

3.9.4 Activities

Based on the project scope and objectives, you need to talk to people in order to identify the sources of knowledge. It can be useful to identify the purpose to which the source of knowledge will be put during the project. These include:

- Interviews: you obviously need to know the names of the people to interview;
- Learning: you may need some written information so you can read it and gain a basic understanding of the domain;
- Extracts: you may need written information from which extracts can be incorporated into your k-base;
- Links: you may need to know the names and locations of documents (electronic and paper). If electronic, you can provide hyperlinks in your k-base (and knowledge web) to the relevant documents. If paper-based, you can provide descriptions in your k-base (and knowledge web) of the contents and locations of the relevant documents.

It can also be useful to classify the type of knowledge sources you are identifying. Here are the main categories:

- **People:** Obviously, most projects rely on named people as a major source of knowledge. These will mainly be the domain experts, but can also include end-users (from whom you can acquire requirements for the end-product);
- **Roles:** It can often be useful to find out what role each of the identified people fulfils. In some situations, you may identify the roles of experts to interview at this stage without knowing their names until later;
- **Organisations and Groups:** It is useful to identify which organisations and groups the identified people/roles work for. You may also find out that certain organisations and groups have knowledge that you need to capture without knowing at this stage who to contact;
- **Information Resources:** Alongside people, you are likely to identify knowledge resources that have already been written by experts. These include all manner of reports, books, manuals, notes, log books, websites and databases. For all but the shortest document, it is important to identify which parts of the document, website or database will be of most use for the project;
- **Locations:** If is useful to identify where the people, organisations, and knowledge resources are located.

It is often a good idea to add the sources you identify into the k-base. This can be done as part of the project description (*i.e.* a page describing where the knowledge was collected) or as actual concepts within the k-base. If you are using an ontology such as GTO (see Appendix 1), then you will be adding concepts to the following classes: People, Roles, Org Units, Info Resources and Locations.

Once you have identified the sources of knowledge, it is useful to create a matrix (grid, table, spreadsheet) showing 'Who knows what' and another showing 'Where is what'. An example of a 'Who knows what' matrix is shown in Figure 2.5 (page 16). An example of a 'Where is what' matrix is shown in Figure 3.5 (overleaf).

		Info Resources					
		Monthly reports	Document WES35	QA Defect Tables	Work Procedures	JICs	Defect Reports
Locations	Company Intranet				☐		
	A-Drive	☐					☐
	B-Drive			☐			
	Company Library		☐		☐		
	Maintenance Database			☐		☐	
	Shop-floor files				☐	☐	☐

Figure 3.5. Example of a 'Where is what' matrix

3.9.5 Problems and Solutions

3.9.5.1 Problem: Fail to Identify Some of the Vital Sources of Knowledge
Sometimes it is not possible to identify all the sources at this stage in a project. This is not a major problem as these will often surface as you go through later steps and especially when you start interviewing experts. Accept the fact that experts know much more than they realise they know and this includes being able to reel off all relevant documents and the names of all other experts. By talking to as many experts as possible in this early stage you are likely to identify most of the sources that you need.

3.9.5.2 Problem: Identify Too Many or Overlapping Sources of Knowledge
The last thing you need at this stage is to be overloaded with a mountain of documents to read through and understand. If a large number of documents are identified, ask the experts which sections of which ones are most appropriate to read. If there are overlapping sources of knowledge, ask the experts (now or during later interviews) which ones are most relevant to be referenced or included in the k-base. Be warned, however, not to treat your k-base as a dumping ground for vast amounts of information already described in other documents and databases. Focus

on identifying in a structured way what already exists, summarising it and providing hyperlinks and locations.

3.9.5.3 Problem: Identify Sources that are Unavailable, Unobtainable
or Inaccessible within the Current Resources, Timeframe and Security Status
It can still be beneficial to identify useful sources of knowledge even if these will not be available to you (for whatever reason) during the project. There are two reasons for this: (i) The sources can be referenced in the k-base so that end-users who can access them know that they exist and their location, (ii) The sources should be documented so that this information is available to anyone extending your project in the future or working on an allied project.

3.9.6 Checklist for Step 9

- Identify all relevant and available sources of knowledge;
- Create matrices of 'Who knows what' and 'Where is what';
- Prioritise all overlapping sources of knowledge

3.10 Step 10: Identify Project Type

3.10.1 Summary

Decide on the type and properties of the project using a 'project description scheme', *i.e.* a list of standard project types and properties. This step should review, build upon and refine the project typing performed in Step 1.

3.10.2 Reasons and Conditions

It is useful during the remainder of the project to determine what sort of project you are doing. This is especially important when: (i) Planning the project; (ii) Deciding if certain steps can be skipped; (iii) Selecting the best way to carry out each step. This step should be performed on all projects unless there is only one methodology supported by the organisation and the project fits exactly into this.

3.10.3 Resources

The resources required for this step are: (i) Project description scheme; (ii) Information captured in Steps 1-9.

3.10.4 Activities

First, you must determine if there is an existing project description scheme available. If there is, you obviously need to obtain it and understand it. If not, then you have to consider developing one that will meet the needs of your organisation.

In developing a project description scheme, you can consider using two main methods:

- A straightforward classification scheme, *i.e.* a set of categories and a way of placing a project into a category;
- A more complex description scheme using a set of properties, *i.e.* describe the project rather than classifying it.

A number of categories, dimensions and properties can be used to classify or describe the project. One, some or all of the following could be used:

- General focus: expert-focused, domain-focused;
- Project length: rapid, short, medium, long;
- Effort from knowledge engineer: 1-day a week, half-time, full-time;
- General depth: shallow and wide, narrow and deep;
- Knowledge depth: basic/explicit, deep/tacit;
- Knowledge focus: conceptual knowledge, procedural knowledge, both;
- Number of experts: none, 1, 2, many;
- Type of experts: academics, professionals, pure performers;
- Role of existing domain documents: core, support, none;
- Purpose: audit, pilot, accelerated learning, end-product;
- End-product: knowledge web, expert system, KBE application, ontology

One way to use a list like this is to select 2 or 3 of the entries and use them to define a classification scheme. A simple, yet powerful, example of this is described below.

3.10.4.1 Classifying a Project Based on Scale

The scale of a project can be derived from two of the factors listed above: the length of the project and the effort from the knowledge engineer. Table 3.3 shows how the scale can be determined.

Table 3.3. Factors that define the project scale as extra-small, small, medium or large

		Length of the remainder of the project		
		4 weeks	12 weeks	24 weeks
Effort from knowledge engineer	**1 day per week**	--	Extra-Small	Small
	2.5 days per week	Extra-Small	Small	Medium
	Full-time	Small	Medium	Large

The four scales shown in Table 3.3 are extra-small, small, medium and large.

- An '**extra-small**' project requires a total effort of 10-12 man days, *i.e.* the knowledge engineer is full-time for 2 weeks or half-time for 4-5 weeks. This would be used for pilot projects and rush jobs (*e.g.* when a person is leaving the organisation in a few weeks time). To do the 47 steps in such a short time would usually require that the majority of Phase 3 is missed out (*i.e.* Steps 30-37 are not performed).

- A '**small**' project requires a total effort of 20-24 man days, *i.e.* the knowledge engineer is full-time for 4 weeks or half-time for 12 weeks. This would be used for small knowledge webs and small ontologies. To do the 47 steps in this time would usually require that some of Phase 3 is missed out (*i.e.* some of Steps 30-37 are not performed).
- A '**medium**' project requires a total effort of 12 man weeks, *i.e.* the knowledge engineer is full-time for 12 weeks or half-time for 24 weeks. This would be used for substantial knowledge webs, large ontologies, knowledge documents and new starters for accelerated learning. All of the 47 steps would be performed.
- A '**large**' project requires a total effort of 24 man weeks, *i.e.* the knowledge engineer is full-time for 24 weeks. This would be used for very substantial knowledge webs and large knowledge documents (*e.g.* for the development of an expert system). All of the 47 steps would be performed, with a particular emphasis on Phase 3 (especially Steps 30, 35 and 36).

This way of classifying a project will be used later when scheduling the project (in Step 12).

3.10.5 Problems and Solutions

3.10.5.1 Problem: The Project Description Scheme is Non-optimal
The way you classify or describe a project does not have to be perfect, but does have to be useful. Do not forget the main purpose of the scheme is to help you plan (hence run) the project. At the very least, the scheme should allow you to select the best procedure to use for the project. For example, during the next step (Step 11) you should be able to answer these questions: How much of the 47-step procedure should I use? What other procedures or methodologies are applicable? Which steps can I miss out or spend less time on? If your scheme for classifying or describing projects does not help answer these questions, then try to amend it so that it can.

3.10.5.2 Problem: The Project is Not Described Correctly
If the scheme is complex or unclear then there is a chance that you will misclassify or poorly describe the project. If this occurs, your choice of procedure and project schedule may lead to problems during your project. There are three solutions here. First, try to use a scheme that is so simple and clear that mistakes cannot be made. The danger, of course, is that if you simplify things too much then the scheme will be of little benefit. Second, seek help from an experienced knowledge engineer or member of the knowledge support team to help you perform the classification or description. Third, monitor your project and how it fits with the procedure and schedule. If you have made an error in classifying or describing the project, then the procedure and schedule will need modifying. Take advantage of any re-planning steps built into your procedure to formally do this (see Steps 21 and 34 for more information).

3.10.6 Checklist for Step 10

- Obtain or define a project description scheme;
- Classify or describe the project using the project description scheme

3.11 Step 11: Define and Understand Procedure

3.11.1 Summary

Define and understand the knowledge acquisition (KA) procedure or methodology to be used. This may require configuring standard procedures or methodologies to fit the exact nature of the project defined in the previous step.

3.11.2 Reasons and Conditions

The project cannot be properly planned if you have not defined and understood the framework and methods to be used to acquire, model and share the knowledge. This step should be performed on all projects. The point at which it is performed may change from project to project. For example, you may want to understand the procedure earlier (perhaps as early as Step 2) so that you conduct the first few steps in the right way. Alternatively, you may want to wait until you know the exact nature and goals of a project before understanding how it will be approached and performed.

3.11.3 Resources

The resources required for this step are: (i) The organisation's KA methodology and procedure/s; (ii) The step-by-step procedure described in this book; (iii) Information on standard methodologies, *e.g.* MOKA and CommonKADS (see Section 7.2); (iv) The project type identified in Step 10.

3.11.4 Activities

In the previous step, you identified what type of project you are doing. You must now match the right procedure with this. Depending on your circumstances, this task may be completely obvious or not at all obvious. Here are some situations:

- You are in an organisation with a single, fixed way of doing a KA project and your project fits completely with the standard way. In this case, the decision is made.
- You are in an organisation with a number of pre-defined KA procedures and a clear way of matching projects to procedures. In this case, the decision should be straightforward.
- You are in an organisation with multiple procedures but no clear way of matching projects to procedures. In this case, you need to find out which procedure suits your project. Input from your knowledge support team will be essential here.

- You are in an organisation with one or more defined procedures but your project does not fit. Now you need to combine elements from different procedures to arrive at one that will suit your project.
- Your organisation has no defined procedures. In this case, you need to find or design one of your own, perhaps based on the procedure presented in this book.

If you need to design a procedure of your own or tweak a standard approach to arrive at one suited to your project, and you (and your organisation) lacks the knowledge to do this, then it may be better to use a more flexible, almost experimental, approach to your project. This would mean trying different things to see which one works the best. This is a risky approach to take and could result in some wastage of time and perhaps upset some of the experts. It also means that planning the project (in Step 12) will be much harder. However, this approach could be worth considering especially if you have a very unique project or have no opportunity to learn a particular procedure.

If you need to merge two procedures to arrive at a composite for your project then select one of them as the main approach and graft onto it aspects of the other one. For example, you may select CommonKADS as your main one and graft onto it aspects of the 47-step procedure (to help with the acquisition of knowledge from the experts). To help you select the right approach, some suggestions are shown in Table 3.4.

Table 3.4. Suggested procedures for particular project aims

Project Aim	Suggested Procedures
Creating a knowledge document for an expert system	CommonKADS plus some aspects of the 47-step procedure (or *vice versa*)
Acquiring knowledge from design engineers for publication on a website	MOKA combined with the 47-step procedure
Acquiring knowledge from design engineers for the development of a Knowledge-Based Engineering application	MOKA plus some aspects of the 47-step procedure
Capturing detailed knowledge of the way an expert solves a specific problem	CommonKADS combined with the 47-step procedure
All other KA projects	The 47-step procedure presented in this book

Whatever procedure or methodology you have selected or been given, you must understand it as much as possible. Hence, get as much training and read as much material as you have time for. Of course, you are already reading this book, which is a good start!

For more information on procedures and methodologies, see Section 7.2 (page 161).

3.11.5 Problems and Solutions

3.11.5.1 Problem: You Select or Define a Procedure that is Inappropriate to the Project or Unachievable within the Project Resources

One way to avoid this problem is to become as knowledgeable as you can about the limitations of the methodologies and procedures that are available to you. This is where reading and learning about different approaches can be valuable. You should also approach the knowledge support team for advice if required. If you do, unknowingly, select a procedure that is non-ideal for the project, then this can affect your approach and will result in difficulties during the project. It is important that you occasionally stand back from the project and ask yourself: "Am I going about this project in the right way?". If your knowledge support team have put in place the right quality procedures then they should have defined points in the project to do this in either an informal or a formal way (*e.g.* using review gates).

3.11.5.2 Problem: You Select or Define a Procedure that is Unclear in Parts

Although experienced knowledge engineers such as myself try to design methodologies and procedures that provide clear and useful advice for every step of a project, there will inevitably be some steps that may be less clear and lead the knowledge engineer to perform tasks in an inappropriate way. The important thing for the subject areas of knowledge engineering and knowledge acquisition is that we knowledge engineers are constantly learning and evolving better methods. This is the reason for the final step in this procedure (Step 47), which is about learning lessons from the project and making suggestions for changes to the organisation's knowledge programme and procedures.

3.11.5.3 Problem: You Select, Define or are Given a Procedure that is Too Difficult to Understand and/or Provides Inadequate Training and Support

Some methodologies are harder than others to understand and apply. In general, the harder ones tend to be those that are more technical in nature and have been written for highly-skilled knowledge engineers. Training courses in KA methods and procedures often need to teach a variety of different skills some of which require a particular frame of mind to fully appreciate and learn. In addition, the knowledge programme may not have the funding and other resources to provide all of the support required by knowledge engineers. This is a real problem because knowledge engineering is not simple and is sometimes under-resourced. So what can be done? Four things. First, try to use software tools that make the KA project as easy as possible. Second, consider using a procedure that is easier to learn and apply even if it is not exactly ideal for the project. Third, actively participate in user groups and knowledge communities (inside and outside your organisation). Fourth, provide suggestions for improvements to the knowledge support team.

3.11.6 Checklist for Step 11

- Select or design an appropriate procedure/methodology;
- Learn how to follow the procedure/methodology;

- Identify any gaps in the procedure/methodology that need to be plugged with ideas from other methodologies or where help from other people is required

3.12 Step 12: Create a Project Schedule

3.12.1 Summary

Create a schedule that says what will be done, by whom and when. The schedule need not be very detailed, but should outline what needs to happen to meet any deadlines or milestones. Complex projects should consider more formal project management methods.

3.12.2 Reasons and Conditions

A schedule is essential so that: (i) People know what to do and when to do it; (ii) Resources are put in place before they are needed; (iii) People, resources and activities are co-ordinated effectively; (iv) Progress can be monitored and assessed; (v) Problems and slippages can be identified early and remedial actions taken. This step should be performed on all projects.

3.12.3 Resources

The resources required for this step are: (i) The procedure/methodology defined in Step 11; (ii) People's availability; (iii) Organisation's key dates; (iv) Project proposal; (v) Project scope;

3.12.4 Activities

As mentioned, complex projects should consider using formal project management methods including specific planning techniques. If you do not adopt such a rigorous approach, then you should at least consider performing the following 6 key activities when creating your project schedule:

1. Identify what sort of schedule you need;
2. Define milestones and review points;
3. Identify constraints (soft and hard);
4. Determine the amount of slack to be built into the schedule;
5. Develop the timescales and required resources;
6. Check and modify the schedule with key individuals

Details of these activities are described below.

3.12.4.1 Identify What Sort of Schedule You Need

Depending on your particular situation, you may need a particular style of schedule. This can be determined by defining what the schedule is going to be used for. Is it to be used to define a day-by-day calendar of events? Is it to be used at a broader level to monitor progress? Is it to be used to manage resources?

3.12.4.2 Define Milestones and Review Points
Many schedules include milestones at key points that have to be reached by certain dates. Project managers often require review points to be held at these milestones so that progress can be monitored. In this way, problems can be identified before they escalate. Key milestones for a KA project might be:

- Project Plan (Step 13);
- Validated taxonomy of relevant expertise (Step 20);
- Validated knowledge base of relevant expertise (Step 29);
- Assessment of prototype deliverable (Step 33);
- Finalised knowledge base (Step 39);
- Release of deliverable (Step 44);
- Review of project (Step 47);

3.12.4.3 Identify Constraints (Soft and Hard);
Constraints are factors that define the boundaries of what can be done. A hard constraint is one that has to be met. For example, you must complete a certain phase by a certain date, or you must see each expert at least twice. Soft constraints are things you aim to satisfy in the schedule but which might not be satisfied if it is not possible. For example, you aim to cross-validate all the knowledge (*i.e.* check it with another expert).

3.12.4.4 Determine the Amount of Slack to Be Built Into the Schedule
The more risks there are with a project, the more slack should be built into the schedule. Only define a tight schedule with no slack if there are few risks. It is not unusual for a very complex project to take 50% longer than originally planned.

3.12.4.5 Develop the Timescales and Required Resources
The timescales will be derived from the procedure that was selected/defined in the previous step (Step 11). If you are using the 47-step procedure, then a simple rule of thumb is as follows: **Spend about 5% of the time available to you on each of the steps**. If this is combined with the simple classification scheme described in Step 10, then the following can be used to create a first-pass schedule:

- For an 'extra-small' project, spend about 2 hours per step, and miss out Steps 30-37;
- For a 'small' project, spend about half-a-day on each step (and possibly skip some or all of Steps 30-37);
- For a 'medium' project, spend between 0.5 or 1.5 days on each step (and perform all of the 47 steps)
- For a 'large' project spend 1-2 days on each step, with particular emphasis on detailed elicitation (Steps 30 and 35) and validation (Steps 29 and 36).

These figures will provide key dates for project milestones as illustrated in Table 3.5.

Table 3.5. Key dates for projects of different scales

Milestone	Step	Extra-Small	Small	Medium	Large
		Scale of Project			
Finalised Project Plan	13	Day 1	Day 1	Week 1	Week 1
Validated taxonomy	20	Day 3	Day 5	Week 3	Week 4
First-Pass knowledge base	29	Day 6	Day 8	Week 5	Week 10
Assessed prototype deliverable	33	N/A	N/A	Week 8	Week 16
Finalised knowledge base	39	Day 8	Day 15	Week 10	Week 20
Release end-product	44	Day 10	Day 19	Week 11	Week 23
Review of project	47	Day 10	Day 20	Week 12	Week 24

To further refine the project schedule, you should do the following:

- Define when to interview the experts, based on the number of experts and the availability of experts;
- Define the time required between interviews, based on the type of transcription (partial or full) and the person doing the transcribing (knowledge engineer or support staff);
- Take account of the number of knowledge engineers on the project;
- Take into account the 'hardness' of the end date (and other deadlines) *e.g.* the day on which an expert is leaving the organisation

3.12.4.6 Check and Modify the Schedule with Key Individuals
Once a provisional schedule is in place, you should check dates with key individuals, particularly domain experts (for interviews) and end-users (for end-product assessment). Feedback from these people will determine what changes need to be made to the schedule.

3.12.5 Problems and Solutions

3.12.5.1 Problem: The Schedule is Deficient on Some Areas, e.g. Does Not Allow Enough Time for Validation
No schedule is perfect and people tend to underestimate the amount of time by ignoring unforeseen circumstances. That said, the schedule should be reasonably accurate if you use the figures and guidelines suggested above. Do not try to think you can do more in less time unless you have the experience to substantiate your claims.

3.12.5.2 Problem: The Schedule is Too Detailed or Lacks Detail
Never try to plan the later stages of a project in too much detail unless you have performed a similar project on many previous occasions and know exactly what

you are doing. If you follow the basic steps set out above, and you are using a methodology that allows you to take stock and re-plan at various stages, then you should be able to cope with most eventualities. If your schedule lacks detail, then you will not know how much you have accomplished at any given point hence you will not know how far behind (or ahead of) the schedule you are. So try to aim for some detail for the next phase and then some key milestones after that.

3.12.5.3 Problem: The Schedule is Too Aggressive and Does Not Account for the Occurrence of Problems that can Slow Things Down and Re-direct Resources
A schedule includes two main things: (i) What to do; and (ii) When to do it. If you do not allow yourself enough time for all the tasks then you run the risk of missing out key steps or rushing things. This may result in errors and mistakes. You might deliver an end-product on time but it may lack the content, structure or presentation to provide maximum benefit to the organisation. An unworkable schedule is almost as bad as no schedule at all. Try to be realistic and always expect the unexpected.

3.12.6 Checklist for Step 12

- Define key project phases and milestones and when they should occur;
- Define time periods when experts are available for interviews and validation sessions;
- Create a schedule that takes account of constraints, dependencies and risks

3.13 Step 13: Collate and Disseminate Project Plan

3.13.1 Summary

Create a project plan by collating information from the project proposal, the project scope and the project schedule. Inform experts, end-users and support staff of the project plan, their roles and their required commitment. Conduct awareness sessions, where necessary, so members of the project team understand the project.

3.13.2 Reasons and Conditions

It is important that everybody involved in the project knows: (i) What the project is aiming to achieve; (ii) What they are expected to do; (iii) When they are expected to do it. If this is not done, then there are the obvious risks that people do not know their project responsibilities and may be unavailable at key stages. This step should be performed on all projects.

3.13.3 Resources

The resources required for this step are: (i) Project proposal; (ii) Project scope; (iii) Project schedule; (iv) List of key people; (v) General awareness and educational material.

3.13.4 Activities

The activities fall into four sections: (i) Create the project plan; (ii) Disseminate the plan to all key people; (iii) Provide general awareness material about knowledge acquisition; (iv) Provide specific material on people's roles and responsibilities.

3.13.4.1 Create the Project Plan
If you have performed the first 12 steps in the procedure, then you should have all the elements needed for the plan. Hence, it is simply a case of collating information from:

- The project proposal (developed and created in Steps 1 to 4);
- The project scope (developed and created in Steps 6 to 8);
- The project schedule (developed and created in Steps 9 to 12)

3.13.4.2 Disseminate the Project Plan
Send the project plan to all those involved in the project. Alternatively, arrange a project kick-off meeting and present the plan using slides.

3.13.4.3 Provide Awareness Information
Awareness information will provide certain individuals with a basic introduction to knowledge acquisition. For example, this can include awareness material for domain experts and the project manager on the procedure and techniques to be adopted on the project and general issues to do with knowledge acquisition within the organisation. This information can be provided on-line, at a specially arranged awareness session or as part of the project kick-off meeting. Whichever method is used, ensure the information is concise, relevant and easy to understand.

3.13.4.4 Provide Role Information
Role information can be provided to help each person to perform his/her role on the project. This can be in the form of Role Sheets, which have generic information for a KA project, supplemented with dates from the project schedule.
The Role Sheet for a **project manager** might include advice on the following:

- Ensure you fully understand the project and its goals;
- Try to keep a watching brief on the progress of the project;
- Try to be available when the knowledge engineer needs help and support;
- Try to be available when major project decisions need to be made;
- Where appropriate, promote the project to other people;

The Role Sheet for a **domain expert** might include advice on the following:

- Ensure you fully understand the project and its goals;
- Try to be available during the following dates to take part in interview and validation sessions...;
- Try to be available during the following dates to validate the knowledge base using web-enabled material... ;
- Be prepared to assist the knowledge engineer in designing plans for interviews (*e.g.* on the questions that will be asked of you);

The Role Sheet for an **end-user** might include advice on the following:

- Try to be available during the following dates to take part in assessment sessions of the prototype end-product...;
- Try to be available during the following dates to take part in assessment sessions of the final end-product...;
- Try to be available during the following dates to attend a launch of the end-product, including a short training course on its structure and usability... ;

3.13.5 Problems and Solutions

3.13.5.1 The Project Plan is Not Sent to All Key Project Personnel and Their Line Managers
The more people know about your project, the higher the chance it will have their support. So send the plan to all domain experts, all end-users and all relevant line managers. Alternatively, provide the plan on-line so that they can read it when necessary, or arrange a kick-off meeting.

3.13.5.2 Project Personnel are Unhappy with the Plan
If you have followed the previous 12 steps, then almost everyone you send the plan to should have been involved in its development. If there are people who object to the plan, then assess their concerns and respond accordingly. Involve the project manager if you need support in convincing people of the project's worth to the organisation. If required, make modifications to the plan.

3.13.5.3 Key Project Personnel are Unavailable or Unwilling to Attend the Kick-off Meeting, or the Awareness Session and/or Read the Project Plan or Role Sheets
People are busy and do not have time to read long, wordy documents or attend long, irrelevant meetings. Hence, to maximise the chances of people reading your plan and attending meetings, make the material as concise, clear and relevant as possible. You should present the plan in a structured way, such as showing the project proposal as a table (see Figure 3.1 on page 30) and the scope as a matrix (see Figure 3.4 on page 42) and a short description. Make sure that meetings last no more than one hour. Give people the information they need and allow time for questions.

3.13.6 Checklist for Step 13

- Collate a project plan and present/send it to the project team;
- Inform individuals of their responsibilities on the project and provide any necessary awareness material;
- Respond to feedback and make any necessary changes to the plan

4

Initial Capture and Modelling

This chapter describes the second phase of the 47-step procedure, which comprises Steps 14 to 29. These steps take the project from elicitation interviews, through knowledge analysis and modelling to validation.

4.1 Step 14: Become Familiar with Basics

4.1.1 Summary

Familiarise yourself with basic knowledge in the domain from the available sources of knowledge (documentation, websites, people). This may or may not involve adding concepts and other content into the k-base.

4.1.2 Reasons and Conditions

It is important to move yourself up the learning curve to prepare for the next steps, *i.e.* preparing and conducting semi-structured interview/s with the experts. The more you can find out now by reading and informal knowledge gathering, the more effective will be the semi-structured interviews and the subsequent steps that involve modelling. This step is highly recommended especially if you are unfamiliar with the domain. If you already know a lot about the domain, then you might want to skip straight to the next step.

4.1.3 Resources

The resources required for this step are: (i) Documents. Books. Websites, Databases; (ii) Experts; (iii) 'Who Knows What' and 'Where is What' matrices.

4.1.4 Activities

The activities here will depend on the project and the availability of information. The essential aim is to get you sufficiently up the learning curve to be able to plan and perform some good semi-structured interviews.

You should already have identified all of the relevant sources of knowledge (during Step 9). You now need to select those that are most applicable to this stage. Here are some of the activities to help you get up the learning curve:

- Read relevant sections from documents, websites and databases. If you require guidance on what to read then consult one or more of your experts. Perhaps arrange a short meeting with experts to do this.
- Analyse and model relevant sections from documents, websites and databases. If the knowledge is already in a structured form, then you might consider adding concepts from these resources into the k-base. If you have no available experts, then this is a must. To do the analysis (and then modelling), use the activities shown in Step 18 and beyond. I advise you not to do analysis and modelling at this stage if:

    - The existing written material lacks structure in the way it describes the knowledge;
    - The existing written material is not focused on your project scope and goals;
    - The project requires structure to be created;

  Note, the analysis and modelling of existing resources with little input from experts can lead to errors in the k-base and a misdirected mindset. It is often better to let ideas form gradually over the next few steps as you interview experts, model the knowledge and conduct validation sessions.
- If there is very little written material, then conduct short (20-30 minute) unstructured interviews or phone conversations with experts and end-users. Focus on getting the basics. Ask very general questions about sub-areas of knowledge, e.g. "What does *semantic web* mean?", "What is it for?" "What knowledge do the end-users need to know about it", "What should I ask the experts?". Note, you may have gathered some of this information during earlier steps. If so, and you feel ready, go to the next step and prepare for a semi-structured interview.
- Attend and observe events such as meetings, presentations, discussions and courses. If the experts and/or end-users attend these information-exchange events, then you might benefit from sitting-in and absorbing some knowledge.
- Start to create a glossary of key terms.
- Start to compile a list of acronyms and their meanings.
- Start to compile a list of things to ask the experts.

4.1.5 Problems and Solutions

4.1.5.1 There are No Available Knowledge Resources to Learn From
It is very unlikely that there will be nothing and nobody available, as you should have identified these resources in Step 9. If there is nothing at all then have a look on the Internet (which has some information on almost everything). If you still cannot locate any available sources of knowledge, and have to wait for experts to become available, then try to talk to some end-users.

4.1.5.2 The Available Knowledge Resources Do Not Cover the Areas of Knowledge Identified in the Scope to the Required Level of Detail
It is worth reading some information even if it is not directly applicable to the project. It will give you a feel for the domain and perhaps some of the jargon that is used. But do not spend too much time reading irrelevant stuff.

4.1.5.3 There is Too Much Material to Read and Not Enough Time, or the Material is Too Difficult to Understand
Do not spend days and days trying to understand things that require a doctorate in the area to comprehend! Pick out terminology and ideas you think are relevant to the project but which you have difficulty understanding. Form these into questions to ask the expert during Step 16. If you cannot understand the material, do not be disheartened. One of the reasons for your project may be to explain this information to others in a clearer and more applicable way. If all the knowledge is already documented in a clear way, then there is little point in doing your project (unless it is to re-structure, amalgamate and re-present the information).

4.1.6 Checklist for Step 14

- Identify the available knowledge resources that can be used to learn the basics;
- Use the knowledge resources to learn the basics;
- Create lists of key terminology and questions to ask the experts

4.2 Step 15: Prepare for Semi-structured Interview/s

4.2.1 Summary

Arrange one or more interviews. Prepare a plan for each semi-structured interview that includes an introduction and questions to ask. Send the plan to the expert/s before the interview/s.

4.2.2 Reasons and Conditions

A semi-structured interview requires a plan that is sent to the expert beforehand and is used during the session. The plan satisfies the requirements described in Chapter 2 in terms of helping the expert to focus on the knowledge required, recall the relevant knowledge and explain things in a clear and logical way. This step must be performed on all projects that involve semi-structured interviews.

4.2.3 Resources

The resources required for this step are: (i) Material and questions developed in Step 14; (ii) Project Scope; (iii) Project Proposal; (iv) Standard question templates; (v) 'Who Knows What' matrix; (vi) Expert/s.

4.2.4 Activities

There are five essential activities in this step:

1. Arrange the interview/s;
2. Identify the knowledge to be acquired during each interview, and summarise it as an introductory section in the interview plan;
3. Write the remainder of the interview plan, which usually consists of a list of questions to ask the expert;
4. Send the plan to the expert at least one day before the interview;
5. Gain feedback from the expert about the plan and make any required changes;
6. Prepare materials for the interview

4.2.4.1 Arrange Interview

When arranging the interview, aim for a one-on-one session with an expert in a meeting room away from his/her place of work. The purpose of this is to avoid unnecessary onlookers and interrupters and to avoid the expert becoming distracted with other people, phone calls, *etc.* Arrange a time and location that is convenient, non-stressful and quiet. Check with the expert that audio-recording is acceptable. Ensure you are seeing the right expert and, if seeing multiple experts, select a sensible order in which to see them. Aim for no more than 3 interviews at this stage, as it is important to start analysing and modelling knowledge before doing too many interviews. A single interview is sufficient for most projects at this stage.

4.2.4.2 Identify the Knowledge to be Acquired

When identifying the knowledge to be acquired, use the information you captured in Steps 9 and 14. If you are unclear about what the expert knows, gain some understanding of the expert's experience and knowledge via a phone call or short chat. Write an introduction using a few short sentences that convey the purpose of the project as a whole (to provide context) and the purpose of the interview.

4.2.4.3 Create the Questions

The number of pre-defined questions to ask will depend on the depth of the knowledge that needs to be explored and the length of the interview. Here are 3 alternatives:

- For a short interview of 20 minutes that covers basic knowledge in not much depth, I would ask 5 or 6 questions (*i.e.* 3 or 4 minutes each);
- For a standard interview of 1 hour which explores issues in greater depth, I would ask about 12 questions (*i.e.* 5 minutes per question);
- For a long interview of more than 1 hour, I would break the session into sections that cover different areas or that tackle the knowledge from different angles. I would use 5 minutes per question, and think about incorporating 1 or 2 paper-based techniques such as a concept map or timeline (see later)

When deciding what questions to ask, there are a few basic ground rules to follow:

- Ask objective questions, not personal ones. Do <u>not</u> ask: "What do you do when you design a widget?" Instead, ask: "What is the best way to design a widget?"
- Ask open questions, *i.e.* do not ask questions that can have a one word answer. Do <u>not</u> ask: "Is designing a widget difficult?", but instead ask: "How difficult is it to design a widget?"
- Ask simple, clear, precise questions. Do <u>not</u> ask: "What are the main tasks, who performs them and the information needed to perform them?". Instead split this into 3 separate questions.

Most questions should start with "What", "How", "Why", "Which" and "Are". Occasionally, questions can start with "Who", "Where" and "When". Here are some examples of good questions:

- "What are the main tasks involved in designing a widget?"
- "What are the basic modelling techniques?"
- "What are the differences between widgets and wodgets?"
- "What types of widgets are there?"
- "How is the stress calculated?"
- "How are the stress results presented?"
- "Why are the stress calculations presented in this format?"
- "Which activities are performed on all the tasks?"
- "Are there any constraints related to manufacturing?"

You can get ideas for questions from a generic model or task templates. If you are using a methodology, such as MOKA or CommonKADS, then the generic k-models described in the methodology can help you to plan and phrase the questions. For example, with MOKA you will be asking questions related to the product model (*i.e.* structural concepts and constraints) and the process model (*i.e.* activities and rules). If you are using CommonKADS then you will have defined (during scoping) which task template you are using. Hence, you will ask questions related to the inferences and the domain knowledge. If you are experienced, you may like to create a meta-model (see Step 24) at this stage and use it to help create the questions.

When you have written the questions, re-read them and for each one ask yourself: "Is this question relevant to the project aims and scope?" and "Will it give me information that can go into the k-base?".

Sometimes you may want to add sections to your interview plan that do not only involve questions and answers. It can be useful to ask the expert to sketch out something on paper during an interview rather than simply talk. This is particularly useful when:

- You are talking about a complex concept that is composed of many other concepts. In this case, it is useful to draw a composition tree on a sheet of paper. An example of a composition tree is Figure 4.8 (page 99).
- You need to know what tasks are performed or what happens during a project. In this case, it is useful to draw a timeline on a large sheet of paper. An example of a timeline is Figure 2.7 (page 18).

- You feel a concept tree can be sketched to show concepts and how they fall into classes. An example of a concept tree is Figure 2.2 (page 14).
- You feel a concept map can be sketched to show how concepts relate to one another An example of a concept map is Figure 2.6 (page 17).

Your interview plan will now be complete. It should look something like the example shown in Figure 4.1.

KBE Project: Interview with Jon Smith (75 minutes)

INTRODUCTION (5 minutes)

The aim of this project is to capture knowledge possessed by expert KBE developers in writing application programs. A small intranet site (approx 80 web pages) will be the main deliverable. The users of the website will be trainee and novice developers. This 75-minute semi-structured interview will cover basic knowledge of the tasks involved in developing a KBE application.

INTERVIEW QUESTIONS (40 minutes)

The following questions will be asked of the expert (approx. 5 minutes per question). Supplementary questions will be asked when required.

1. What are the <u>main</u> tasks involved in writing an application?
2. For each of the main tasks, what information is required?
3. For each of the main tasks, what is created or achieved?
4. What typical bugs occur?
5. For each of the typical bugs, what debugging activities take place?
6. What different types of KBE application are there?
7. How do the different types affect the development process?
8. Given the remit of this interview, is there any important information that we have not yet covered?

TIMELINE (30 minutes)

Using paper and pencil, we will draw a timeline of the KBE development process from start to finish. The timeline will show all of the main tasks, and when they occur. The dependences between tasks will be discussed. We will also cover the conditions under which certain tasks are performed in a different order or more than once.

Figure 4.1. Example of a plan for a semi-structured interview

After you have created your interview plan, it is a good idea to get the opinion of another person, such as an experienced knowledge engineer.

4.2.4.4 Send the Plan

You should now send the plan to the expert. This should be at least 1 day before the interview is to take place. This will give the expert a chance to read the questions and provide any feedback.

4.2.4.5 Gain Feedback

Phone the expert later to ask for comments and any changes. You can ask questions such as: "Do the questions address the project scope?"; "Do the questions make sense?"; "Can you answer all the questions?"; "How can the questions be improved or re-worded?"; "What questions are missing?". Based on the expert's feedback, make any modifications that are required.

4.2.4.6 Prepare Materials

Finally, prepare any resources and materials that you may need at the interview. Always remember to take the following to the interview:

- Audio- or video-recording equipment;
- Paper and pencil (for the expert to sketch diagrams or make notes);
- Two printouts of your interview plan (one for you and one for the expert)

4.2.5 Problems and Solutions

4.2.5.1 Questions Do Not Address the Scope

From Steps 6-8, you should have identified and agreed a scope for the project, *i.e.* the areas of knowledge to acquire. Do not forget that the scope defines the boundaries and subject areas to be covered during the interview. Try to ensure your questions cover areas within the scope. The expert can help you do this if you: (i) Include a summary of the scope as part of your interview plan; (ii) Ask the expert prior to the interview if the questions fit the scope.

4.2.5.2 Questions are Badly Expressed or Confusing to the Expert

Be clear about what you want to ask and then form each question in a clear way. Use the examples given previously. It is usually a good idea to start a question with 'how', 'what', 'why', *etc.* Make sure you ask one thing per question. When you check the interview plan with the expert, ask "Do you understand all of the questions?"

4.2.5.3 There are Too Few or Too Many Questions

As a rough rule of thumb, 4 or 5 minutes per question is usually adequate. If you allow less than 3 minutes per question (*e.g.* 20 questions in 1 hour) then a number of problems can arise: (i) The expert may feel constrained and pressured to answer in such a short time and not explain important points, (ii) You run out of time and cannot ask all the questions, (iii) You spend much longer with the expert than planned which is unprofessional and may discourage the expert from future sessions. Of course, if the questions are very simple and direct (which is rare but sometimes the case), then less than 3 minutes may be fine. If you allow more than 7 minutes per question then you may be putting undue emphasis on certain issues and going into too much detail that is not required at this stage. Remember, you will be seeing the expert again to validate the knowledge and again (if required) to gather more detailed knowledge.

4.2.6 Checklist for Step 15

- Create an interview plan that includes an introduction and a set of clear questions;
- Send the interview plan to the expert before the interview;
- Obtain feedback from the expert on the interview plan

4.3 Step 16: Conduct Semi-structured Interview/s

4.3.1 Summary

Ask questions of an expert using the interview plan developed in Step 15. Ask supplementary questions to clarify and probe what the expert says.

4.3.2 Reasons and Conditions

The semi-structured interview technique is designed to embody the four major aims of knowledge elicitation, *i.e.* Focus, Recall, Explain, Check (see below for more details). As such, it an essential method for gathering knowledge from experts in an efficient and effective manner. This step must be performed on all projects that involve one or more experts.

4.3.3 Resources

The resources required for this step are: (i) Interview plan; (ii) Recording equipment; (iii) Pen and paper; (iv) Expert; (v) Meeting room.

4.3.4 Activities

The key activities at a semi-structured interview are as follow:

- Ensure the expert is comfortable and happy with the structure and content of the interview plan;
- Set out the aims and context for the interview using the introduction that you wrote on the plan;
- Check that the audio/video recording equipment is working and set it going;
- Go through the interview plan, asking each of the pre-defined questions;
- Listen carefully to what the expert says and do not write many (if any) notes;
- Ask extra, unplanned questions in order to clarify what the expert has said (*e.g.* "What does RDF stand for?");
- Ask extra, unplanned questions in order to probe for more detailed information (*e.g.* "Could you say a little more about mould design?")
- Make succinct summaries back to the expert to verify that you have understood what was said;

- Ask the expert to explain again if you are getting completely lost, *e.g.* say something like: "I'm sorry but you've lost me there. Could you describe the use of Coplanar PCBs again?";
- Encourage the expert to explain domain-specific terminology and acronyms;
- Encourage the expert to use paper and pen when necessary (especially if the subject is very technical and/or the expert is using a lot of hand gestures to try to draw diagrams in the air);
- Politely steer or interrupt the expert if he/she is straying from the scope of the project, *e.g.* say something like: "I'm sorry but I think we're straying a little from the question. Perhaps we could move on to the next question?"
- Try to get the right balance between: (i) Listening and waiting, and (ii) Prompting, steering and suggesting;
- Identify any gaps in the expert's knowledge and where to find that knowledge

These activities are aimed at helping the expert to focus on the project scope, recall all the relevant knowledge and explain in a clear manner. The activities should also help you to understand what is said and check that you have understood it. Here are the reasons for this:

- Helping the expert to **focus** on the scope is achieved with the introduction, the questions, careful steering and polite interruptions;
- Helping the expert to **recall** the knowledge is achieved with the expert seeing the questions beforehand, careful prompting, waiting and listening;
- Helping the expert to **explain** is achieved with the interview plan, careful prompting, use of paper and pen, and any props;
- Helping to **check** the knowledge is achieved with summaries back to the expert and the noting down of key terminology for a glossary

The use of an audio or video recording has a number of benefits:

- It captures everything the expert says so nothing is missed;
- It removes the need to scribble notes which distracts you from listening;
- It is a perfect record of what was said, as opposed to scribbled notes which can miss vital knowledge and can be misinterpreted afterwards;
- It is an important knowledge resource afterwards (*e.g.* for end-users to listen to audio clips or view video clips on a website);
- It helps make analysis and modelling easier and more effective

If you cannot make an audio recording, then the next best alternative is to ask a colleague to sit-in on the interview and write/type notes.

At the end of the interview, you should walk away feeling that a lot of good ground was covered and that you understand a lot more about the domain. You should have plenty of new concepts and descriptions to add to the k-base. The expert should leave the room feeling that he/she has had the time and space to describe his/her knowledge.

4.3.5 Problems and Solutions

4.3.5.1 The Expert Says Too Much, Says Too Little, Does Not Explain Clearly or Does Not Stick to the Scope

It is your job to help the expert to focus on each question and answer it as clearly as possible. The main ways to do this are use of the interview plan and the way you behave at the interview. If the expert is being unclear, say "sorry, I'm not getting this, could you please describe it again or from another angle". Sometimes it is necessary to interrupt the expert. If so, say "sorry, I think this is getting into too much detail, I wonder if we could come back to the question". There is sometimes a balance to be struck between: (i) Knowing what you want to try to get out of the expert, and (ii) Letting the expert dictate the ground to be covered. Too much emphasis on the former and you will constrain (and perhaps frustrate) the expert. Too much on the latter and you might drift from the scope and waste valuable time. Always keep asking yourself: "Is this the sort of stuff I want to put into the k-base?" and "Will my end-users find this stuff useful?".

4.3.5.2 There are Onlookers Who Interrupt and Distract the Expert

When arranging the interview, try to avoid any onlookers. A one-on-one session is usually best. If you cannot avoid the presence of other people, then try to set the ground rules before the interview begins. An important ground rule will be the times at which people can make comments and ask questions. For example, you could appoint a chairperson who will ask for questions and comments at the appropriate times, such as at the end of each numbered question.

4.3.5.3 You Cannot Think of Good Supplementary Questions

Some people have a natural aptitude for asking good, probing questions and others do not. If you cannot think of good supplementary questions during the interview, then try one of these: (i) Pick-up on a piece of terminology that you do not fully understand and ask the expert to explain it; (ii) Summarise back to the expert what has been said and ask if it is correct; (iii) Ask for an example or illustration; (iv) Do not say anything – sometimes it is better to move on to the next planned question than ask redundant and unnecessary supplementary questions. Try to ensure that you do not ask questions for the sake of it, for example those that you personally find interesting if these are not within the scope. Aim for 'must-haves' rather than 'nice-to-haves'.

4.3.6 Checklist for Step 16

- Conduct the semi-structured interview in the right manner;
- Cover the ground that was required to be covered;
- Obtain a good audio- or video-recording of the interview

4.4 Step 17: Transcribe Interview Recordings

4.4.1 Summary

Listen to the audio recordings made during the interview/s and create notes or a full transcript.

4.4.2 Reasons and Conditions

For most projects, the words spoken by the expert are the most important source of knowledge to dissect and rearrange when creating the k-base. Having a written record of some or all of what the expert says makes the next few steps of analysis and modelling much easier and much more effective. This step need not be performed when: (i) There is no audio recording; (ii) The project is short and involves an experienced knowledge engineer (where analysis and modelling can be done directly from the audio recording).

4.4.3 Resources

The resources required for this step are: (i) Audio tapes or files; (ii) Play-back equipment and/or software; (iii) Someone to do the transcribing; (iv) Word-processor software or a specialised tool (perhaps with voice recognition software).

4.4.4 Activities

Firstly you must decide how the transcription is to be done and by whom. Here are the alternatives:

- The knowledge engineer does a full transcription, *i.e.* types everything that was said at the interview or speaks it into a voice recognition system;
- A secretary does a full transcription;
- The knowledge engineer does a partial transcription, *i.e.* types selected portions of the interview;
- The knowledge engineer identifies concepts, *i.e.* types key words/phrases

Note, that the secretary cannot do the 3rd and 4th alternatives, as he/she will not know enough about the project or domain to select the key sections and words.

The most important factor to bear in mind is that a full transcription can take a long time if you are typing. For every 1 hour of audio recording, an experienced secretary will be take around 5 or 6 hours to type this into a document. This can be less if the expert speaks slowly and clearly, but can be more if the expert speaks quickly, uses a lot of jargon and the quality of the audio recording is poor.

Although use of voice recognition technology is an option (see later), most transcripts are typed by hand. So what option should one choose? Here are some factors to consider:

- Is there a secretary available, willing and able to do a full transcription?
- Can the secretary do the transcribing in the next few days?
- Does the knowledge engineer have the time to do a full transcription?
- Would a partial transcription be better than a full one?

To help you make this decision, it is worth giving you some advantages and disadvantages of each option. These are shown in Table 4.1.

Table 4.1. Advantages and disadvantages of transcribing options

Option	Advantage	Disadvantage
Secretary does a full transcription	The knowledge engineer can get on with other things There is a complete record of what the expert said, sections of which can be copy-and-pasted into k-pages	The transcript can contain many mistakes The knowledge engineer does not re-live the interview
Knowledge engineer does a full transcription	The knowledge engineer re-lives the interview There is a complete record of what the expert said, sections of which can be copy-and-pasted into k-pages There will be few mistakes	Takes the knowledge engineer away from other tasks Can be hard work, especially if you are not a good typist
Knowledge engineer does a partial transcription	Saves time (at this stage) The knowledge engineer partially re-lives the interview	Incomplete record of what the expert said May need more work later when adding content to k-pages
Knowledge engineer selects only concepts	Saves time (at this stage) The knowledge engineer partially re-lives the interview	Limited record of the interview Will need more work later when adding information to k-pages

Whichever method you use, here are some hints and tips:

- Do not listen to too many words before typing them. It saves time (in rewinding and listening again) if you can stop the recording after a few words, finish typing them, then listen to the next phrase that the expert says;
- Use the interview plan so that you do not have to type the questions that you read out at the interview;
- The expert will probably use a lot of "ums" and "errs". Do not type these.
- The expert may start a sentence then pause and repeat the same words or change direction. Do not type fragments that contain no meaning.
- If the expert stops halfway through a sentence and changes direction, put a hyphen (dash) to show this;
- If there are interruptions and sentences are not completed, use dots, like this …
- Have the questions and comments made by the knowledge engineer in bold so that is it easy to see what the expert said;

- Do not put the names of the people involved before every section of speech. If more than one expert is present, use a numbering scheme. If more than one knowledge engineer is present, then ignore which one spoke what (just put them all in bold).
- If the expert is referring to a document or sketch, there may be "this" and "that's" that are ambiguous. If so, it can be useful to put in square brackets what the expert was pointing/referring to.
- If someone else is doing the transcribing, give him/her a list of the domain-specific terms that were used by the expert at the interview. This is very important and can save a lot of time later in correcting the transcript.

As mentioned, the use of voice recognition software is an alternative to typing. Current technology is not too good: it requires a lot of time and effort to train for a particular person's voice, and there needs to be a significant amount of correcting afterwards. That said it is worth considering this option as it can vastly reduce the time taken. Voice recognition can be used in a number of ways:

- A secretary, knowledge engineer or another person listens to the interview recording on headphones and simultaneously speaks what is being said into a microphone connected to the voice recognition system;
- The knowledge engineer listens to a section of the recording then stops it and summarises what the expert says by speaking into the voice recognition system;
- The previous option is done at the actual interview, *i.e.* the knowledge engineer takes time out at the interview to summarise to the expert what was said and records this for later voice recognition;
- The audio recording of the interview is plugged directly into a voice recognition system that has been trained to recognise the voice of the expert. With current technology, this training requires around 2 hours. Hence, this is only an option if you have one expert and have a lot of availability from this person

In the future, when voice recognition technology improves, there is likely to be no need to train a system to a particular person's voice. It will simply be a case of capturing the interview directly into the voice recognition system and walking away from the interview with a full and perfect transcription.

4.4.5 Problems and Solutions

4.4.5.1 There is Not Enough Time or Resources to Do the Transcribing
In an ideal world, the audio recording of the interview will be fully transcribed by someone other than you and be complete within a day of the interview. Unfortunately, this is unrealistic for many projects. But things are not so gloomy. It can be beneficial to do-it-yourself and to have a partial rather than a full transcript. By doing it yourself, you can re-live the interview experience, which can be more rewarding than just reading an interview transcript. By listening through, the knowledge becomes more deeply embedded in your mind. Also, you learn how to be a better interviewer by hearing the mistakes you make. You (like I have done) will ask yourself "why didn't I ask that?" and "what made me say such a stupid

thing?" An advantage of a full transcript is that you have a complete record and can cut-and-paste sections of text into the k-base. An advantage of a partial transcription is that you have started to edit and analyse what the expert said, thus saving time. Also, you will be less inclined to add too many irrelevant concepts into the k-base when you come to Step 18.

4.4.5.2 The Audio Recording is Very Poor Quality

As a knowledge engineer, you often learn things by bitter experience. Simple things like having spare batteries, checking the recording is working, making sure that the room is quiet, putting the microphone close to the expert, are the things you learn once you have made a bad recording. A specialised digital recorder or a laptop can be useful as you can keep an eye on the sound level and easily copy-and-send files to other people. Using a decent microphone is important. Do not use cheap equipment that gives poor quality whatever you do. If it is hard to understand the recording, do a partial transcription yourself as soon after the interview as possible while you remember what was said.

4.4.5.3 The Transcription Contains Many Mistakes or Ambiguities

A major disadvantage of having someone else do the transcription is that they probably do not know the jargon and terminology used by the expert. This can result in some amusing mistakes and misinterpretations of the words the expert used. This can require a lot of correcting by you. The best way to avoid this problem is to send a list of the terminology and acronyms used at the interview to the transcriber. This will lead to a better transcript and a less onerous task for the transcriber.

4.4.6 Checklist for Step 17

- Decide whether a partial or full transcription is required;
- Produce an accurate and useful record of what the expert said;
- Save the transcript file to an accessible location

4.5 Step 18: Perform Initial Analysis

4.5.1 Summary

Identify concepts from interview transcripts and/or other domain documents, categorise them, and add them to the k-base. If required, also identify attributes, values and relations.

4.5.2 Reasons and Conditions

The main components of a k-base are concepts. This step is essential as it adds concepts into the k-base. This step must be performed on all projects.

4.5.3 Resources

The resources required for this step are: (i) Interview transcript/s; (ii) Documents; (iii) Tool that supports knowledge analysis.

4.5.4 Activities

Knowledge analysis was introduced in Section 2.2 (page 12). It is usually performed with a set of marker (a.k.a. highlighter) pens on a paper document or simulated marker pens in a software tool. Each pen represents a different class of elements in the k-base, *i.e.* a concept, a sub-class of concepts, an attribute, a value or a relation. Each main class will have a marker pen with a specific colour.

An example of part of an analysed transcript is shown in Figure 4.2. Due to the monochrome nature of Figure 4.2, you cannot see the use of colours. In the original screenshot (from PCPACK) each of the 17 pens has a different colour.

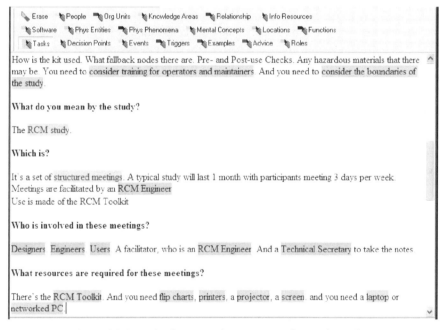

Figure 4.2. Use of software marker pens to analyse a piece of text

Marking-up achieves two things:

- By marking up (highlighting) a word or phrase with a marker pen, this identifies that word/phrase as being an element that should be included in the k-base;
- The particular marker pen that is used will determine the class to which the element belongs. Hence, an '*is a*' relation should be added between the element and the class in the k-base.

For example, if I mark-up the phrase "Design the rear bumper" with a marker pen representing 'Task', then:

- 'Design the rear bumper' should be added to the k-base;
- The relationship 'Design the rear bumper – is a – Task' should be added to the k-base

If you are using a specialised tool such as PCPACK, then this will be done automatically. If not, you may have to do this by hand.

So, there are two main activities during knowledge analysis: (i) Selecting the marker pens to use; (ii) Going through the document, marking-up words and phrases. Let us look at each of these activities in more detail.

4.5.4.1 Selecting the Marker Pens
For most projects the majority of the marker pens will be generic types of concept (see Section 2.2.1 on page 12 for some examples). In some cases, domain-specific maker pens will be used.

The marker pens that you select could be taken from a generic ontology (such as GTO), a general methodology (such as MOKA), a problem-solving model (such as in CommonKADS) or created specifically for your project.

It is vital during analysis to have the right set of marker pens and to know what each one means. As a general rule, you should pick marker pens that represent classes that were talked about during the interview. Here are some ideas:

- If the expert talked about a whole range of different things, then use a single marker pen that represents 'Concepts'. By marking every relevant concept as this, you can sort and categorise them into classes when you construct your concept tree (in Step 19).
- If the expert talked about a small set of general things, then use these as marker pens. For instance, the classes 'Physical Entities', 'Documents', 'People', and 'Tasks' can cover many domains.
- If the expert talked about a specific set of things, e.g. pesticides, or whales, or business issues, or cameras, then use these as marker pens. However, do not make the mistake of marking-up the components of something using the same pen, e.g. marking 'camera lens' with the 'camera' pen is incorrect (as it is not a type of camera), but use a 'camera components' pen instead.
- If you want a general set of marker pens that cover most domains, then you can use the 17 marker pens defined in GTO (see Appendix 1). These are the marker pens shown in Figure 4.2.
- If you are using a specific methodology, then use the classes in it as pens. For example, using MOKA your pens would be 'structural entities', 'constraints', 'functions', 'activities', 'rules' and 'illustrations'. If you are using the diagnosis model in CommonKADS, then your marker pens might be 'symptoms', 'hypotheses' and 'diagnostic techniques'.

If you are using GTO, then you may have already identified a number of concepts during Steps 6 and 9. These are Knowledge Areas and/or Tasks (in Step 6), and People, Roles, Org Units, Locations and Knowledge Resources (in Step 9). Hence, the main focus in this step is to identify: Phys Concepts, Software, Mental

Concepts, Knowledge Areas, Tasks and Events. You might also mark-up Phys Phenomena and Functions if the project is focused on a complex physical concept (*e.g.* the design, manufacture, maintenance or use of a product or component). It is probably better not to mark-up the remaining concepts (Advice, Examples, Triggers and Decision Points) as these are better added during modelling.

4.5.4.2 Using the Marker Pens

One of the biggest mistakes made by many knowledge engineers is too much marking up, *i.e.* identifying too many concepts. PLEASE DO NOT MAKE THIS MISTAKE – DO NOT IDENTIFY TOO MANY CONCEPTS!

So how many concepts should you identify? This depends on the size of the project. Roughly speaking, if the deliverable is a knowledge web then by the end of the project you should have identified about 10 concepts for each day of effort from the knowledge engineer. For example, for a project of 20 days of effort from the knowledge engineer, then about 200 individual concepts should be identified, although not all of these will be identified at this initial stage of analysis as many will be added during later capture steps.

If your project is to create an ontology, then the number of concepts can be much larger, as you are likely to be describing each one in less detail.

4.5.5 Problems and Solutions

4.5.5.1 Difficulties in Classifying Concepts

This can be a problem even for experienced knowledge engineers. A generic ontology such as GTO can be useful as it defines the high-level classes to identify in relatively understandable terms. If you are using a dedicated KA tool (such as PCPACK), then it will not matter if you get things too wrong as it will be easy to re-classify concepts later. If you are struggling, it is better to get the concepts into the k-base then try to work out what class they are (when you create and validate the concept tree in Steps 19 and 20). If you are completely unsure, just identify the concepts as a very high-level class (such as 'concepts') and leave the classification until later.

4.5.5.2 Identify Too Many Concepts

As already said, this is a common mistake made by many knowledge engineers. Unless you are creating an ontology, it is very unlikely that you will be trying to pick out everything that can be a concept and putting it into the k-base. You must be more selective. If your project is to produce a knowledge web, then ask yourself "do I want this as a page in my website?" If the answer is no, then do not select the word or phrase as a concept. Do not fall into the trap of thinking that concepts describe all of the knowledge. They do not. They are the basic structure with which to describe the knowledge. Here is a useful analogy to think about: concepts are like the branches of a Christmas tree, and the other contents of the k-base are like the baubles, tinsel and other decorations. In other words, you will add the extra knowledge contained in the transcript to the k-pages directly as descriptions without formally modelling it in the k-base. Remember, also, that you can (and will) be adding more concepts later when you see the expert and when you use k-models to edit and add to the k-base.

4.5.6 Checklist for Step 18

- Identify an appropriate number of concepts;
- Identify each concept as belonging to a major class;
- Add the concepts to the k-base

4.6 Step 19: Create Concept Tree

4.6.1 Summary

Create a hierarchical diagram of concepts showing classes and class members.

4.6.2 Reasons and Conditions

A fundamental aspect of knowledge is that of classes and types. A concept tree is a simple hierarchical diagram that shows classes and types, as was described in Section 2.3.1 (page 14). It is an essential part of knowledge acquisition to define a taxonomy, *i.e.* to define the class to which a concept belongs and to create concepts that represent classes. This step must be performed on all projects.

4.6.3 Resources

The resources required for this step are: (i) A k-modelling tool that supports the creation of a concept tree; (ii) Concepts identified in Step 18.

4.6.4 Activities

There are 4 basic activities when creating a concept tree:

- Moving nodes (concepts) around the tree;
- Adding new nodes (concepts) to a tree;
- Deleting nodes (concepts) from the tree;
- Renaming nodes (concepts) in the tree

This is relatively straightforward even if you are not using a specialised tool. The difficult part is avoiding the problems that can occur. To do this requires knowing:

- That all the links on a concept tree represent the 'is a' relationship;
- The terminology that is used to describe trees;
- What classes to use in a tree;
- The naming conventions to use;
- How to deal with complex cases, *e.g.* multiple parents and synonyms

Let us look at each of these 5 areas in turn.

4.6.4.1 Using 'Is A' Relationships

The number one rule when creating a concept tree is to ensure that all the links represent 'is a'. Remember that 'is a' is equivalent to 'is a type of'. To test your ability to do this, look at the tree shown in Figure 4.3 and try to spot the errors in it.

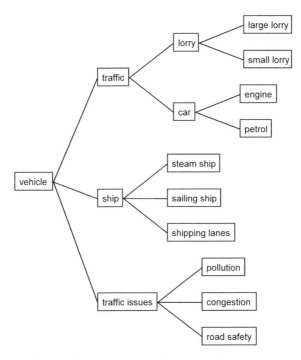

Figure 4.3. Concept tree with some deliberate errors

How did you do? I hope you spotted that the errors are: (i) *traffic* is not a type of *vehicle*, it is more of a synonym (hence should be deleted as a concept or placed in a 'synonyms' category); (ii) *engine* is not a type of *car*, hence should be placed somewhere else, perhaps in a 'vehicle components' category; (iii) *petrol* is not a type of *car*, hence should be placed somewhere else, perhaps in a 'vehicle requirements' category; (iv) *shipping lanes* are not a type of *ship*, hence should be moved somewhere else, or perhaps deleted (and described in the k-page for ship); (v) *traffic issues* are not a type of *vehicle* so should be moved. Note that many of these corrections would require a 'concepts' node at the top (root) of the tree rather than the more specific node 'vehicle'.

4.6.4.2 Terminology

Let me now explain some of the terminology used to describe a tree. There are two analogies: actual trees and family trees. From actual trees we get the following:

- Root node (or 'root' for short) is the highest-level node in a tree, *e.g.* 'vehicle' in Figure 4.3.

- Leaf nodes are at the ends of a tree, *i.e.* nodes without sub-types. For example, 'steam ship' and 'congestion' are leaf nodes in Figure 4.3.
- Branch is one part of the tree going from the root to leaves, *e.g.* there are three branches in Figure 4.3 (for traffic, ship and traffic issues)

From family trees we get the following:

- Parent is the node one level higher than the one being referred to, *e.g.* 'ship' is the parent of 'sailing ship' in Figure 4.3.
- Children, or child nodes, are the nodes one level lower than the one being referred to, *e.g.* 'sailing ship' is a child of 'ship' in Figure 4.3.
- Descendents are all of the children and children's children of a node.

4.6.4.3 Classes to Use in a Concept Tree

A class is a concept that has children on a concept tree, *i.e.* other concepts are related to it by the 'is a' relation. Let us now consider what classes to use in a tree. This is rarely a simple matter, as the classes tend to evolve as the project progresses. You may have already defined some classes by creating and using marker pens in the previous step. In developing further classes, it can be a case of using a top-down approach, a bottom-up approach or a combination of the two.

- Using a top-down approach, you start with a general set of classes (*e.g.* those included in GTO) and work down the tree adding sub-classes where necessary.
- Using a bottom-down approach you start with no high-level classes (other than something very abstract such as 'concept') and develop classes by: (i) Grouping concepts that are similar; (ii) Giving names to the groups; (iii) Grouping the groups; (iv) Naming the super-groups, and so on.

4.6.4.4 Naming Conventions

It is important that the names of all concepts in the tree follow standard naming conventions. These are:

- Start tasks, and only tasks, with an active verb (*e.g.* make, design, perform)
- Do not worry about using plurals. A purest would tell you to put 'vehicle' rather than 'vehicles', but it rarely makes much difference. Do what seems right. My preference is to use: (i) Plurals for the names of classes, *e.g.* 'ships'; (ii) Singulars for leaf nodes, *e.g.* 'steam ship'.
- Make names as clear and precise as possible. The name should be able to be read and understood away from its context. For example, you may have 'petrol' and 'diesel' as child nodes of 'cars', meaning 'petrol cars' and 'diesel cars'. If so, then give them the full name to distinguish them from the names of fuels. If not, then have them in another place, such as under 'fuels'.
- Do not have two concepts with the same name but with different meanings, *e.g.* 'shipping' meaning things that float on the water, and 'shipping' meaning the delivery of goods from a factory. If you cannot find a suitable

synonym for either one, then consider placing the context in brackets, *e.g.* 'shipping (ships)' and 'shipping (delivering)'.

- Do not have two concepts with different names that mean the same thing (*e.g.* 'vehicles' and 'traffic' in the above example). Choose one of them, and add the other either as part of the description in a k-page or in a special 'synonyms' class (and use the 'has synonym' relation to link them).

4.6.4.5 Complex Cases
Sometimes it can be difficult to define a class as one thing and not another. Suppose the expert talks about the following in an interview "large lorries", "small lorries", "fast lorries" and "slow lorries". The expert then describes a particular lorry as being large and fast and another as being small and slow, *etc.* How do you represent this in a concept tree? The answer is that you can, but rarely should. There are three ways to deal with a situation like this:

- Show the information in the concept tree with multiple parents;
- Select one set of classes ('large lorries' and 'small lorries', or 'fast lorries' and 'slow lorries') for the concept tree and show the other information in another way (using attributes/values or as descriptions in a k-page)
- Have no sub-classes of lorries in the concept tree and describe them using attributes/values or as descriptions in a k-page

Sometimes there are different ways of structuring the concept tree. This usually means deciding what sub-classes to use. For example, suppose I have 'whiskey'. Do I classify it as 'alcoholic drink', 'expensive drink' or 'grain-based drink'? The choice usually depends on two factors: (i) The project scope, and (ii) The way the expert thinks of it. If in doubt, leave the tree flat (without many classes and sub-classes) and ask the expert in Step 20 to provide some classes.

4.6.5 Problems and Solutions

4.6.5.1 Some of the Relationships are Wrong, i.e. are Not 'Is A'
There is a simple trick to check if the relationships in your concept tree are all 'is a'. Look at each of the leaf nodes (those at the end of the tree with no children) and compare it to the high-level class that it belongs to towards the 'top' of the tree. Ask yourself "is this (the leaf node) a type of this (the high-level class)?" If the answer is no, then one or more of the intervening relationships is not 'is a'. For example, suppose you have 'piston' as a leaf node under the high-level class 'vehicles'. Is a piston a type of vehicle? No, it is part of a vehicle. Something is wrong. You can detect where the problem is by repeating the procedure for the parent of the leaf node and trying again. When you have located the problem, either: (i) Move the offending node or branch to another place (*e.g.* from under 'vehicles' to under 'vehicle components'); (ii) Rename the concept; (iii) Add a new class and then move the offending node or branch to it.

4.6.5.2 The Tree is Too Flat or Too Bushy
There is an art to getting a well-structured concept tree that a knowledge engineer gains with experience. Roughly speaking, try to have between 2 and 10 child nodes

directly under each class or sub-class. Having nodes with one child is fine, if you think the expert will add more when you next see him/her (in Step 20), but not as a final way of having the tree. Except for certain types of concepts (*e.g.* tasks), you should have some sub-classes and sub-sub-classes for most high-level classes.

4.6.5.3 Not Re-naming Concepts to Have Clear and Standardised Names

The names of concepts should be clear, unambiguous and comprehensible if read on their own. If the name is ambiguous (has more than one meaning within the subject area of your project), then you should add something to the name to disambiguate it. For example, if you have two concepts both called 'case', then you should rename them to be, say, 'case (container)' and 'case (legal action)'. Alternatively, you might rename the first 'container' (if that is a synonym) and/or the second as 'legal case'. Sometimes you may have concepts whose name relies on their position in a k-model to fully convey the meaning. For example, in a concept tree you may have types of a chair shown as 'swivel', 'reclining' and 'collapsible'. Read on their own, it is not obvious that these are types of chair. It is much better to be pedantic and call these concepts 'swivel chair', 'reclining chair' and 'collapsible chair'. A common mistake made by many knowledge engineers is not to use the standard naming scheme for tasks, activities and processes. These should always start with a verb and have a word or phrase after the verb. You should check the name by putting the word 'They' in from of it and seeing if it makes sense. For example, the following phrases make sense: "*They* design the spars", "*They* conduct the interviews", "*They* allocate the resources", hence these are good names. "*They* spar design", "*They* interviewing", "*They* allocating resources" do not make sense, hence are not good names. If in doubt about names, wait until you see the expert in Step 20 and ask him/her to check that you have clear and sensible names for each of the concepts.

4.6.6 Checklist for Step 19

- Ensure all the concepts are in the concept tree and all have a good name;
- Ensure all branches are checked for 'is a' consistency;
- Ensure new concepts (representing classes) are added to the tree so it shows a taxonomy of the relevant domain knowledge

4.7 Step 20: Validate Concept Tree

4.7.1 Summary

Check the concept tree with one or more experts and make any changes or additions that create a more useful and knowledge-rich concept tree.

4.7.2 Reasons and Conditions

As with all k-models, the expert/s need to check that the information displayed is correct, relevant and complete. It is important to use this session to clarify any misunderstandings and terminological difficulties. This step must be performed on all projects (unless there is no expert involved in the project).

4.7.3 Resources

The resources required for this step are: (i) Concept tree on paper or on a computer screen; (ii) A knowledge modelling tool; (iii) Expert/s; (iv) Meeting room.

4.7.4 Activities

Before the validation session, you should do the following:

- Decide which experts to consult if more than one was interviewed. You will usually, at this stage, validate knowledge with the same expert that you acquired it from. Later, in Step 36, you can do cross-validation, *i.e.* show knowledge captured from one expert to another expert.
- Decide how long the session will take. I suggest you allow at least 30 seconds for each node (*i.e.* concept) in the tree. For example, you need about 1 hour to check a tree with 100 nodes.
- Decide on the order in which to present parts of the concept tree.
- Decide whether to use a paper printout of the tree or display it on a computer screen. If you are using specialised software (such as PCPACK), then you can edit it directly at the session thus minimising mistakes and reducing time afterwards.
- Decide whether to use a face-to-face session or do the validation remotely. I recommended that you to do a face-to-face session. However, if this is not possible or feasible, then the session can be performed remotely. A good web-enabled meeting/conference application is ideal for this. Although it is not recommended, there are times when you may have to do the validation by sending the concept tree via email or post with: (i) A feedback form (ii) A way for the expert to edit the tree, (iii) A follow-up phone call.
- Arrange the session with the expert. If it is a face-to-face session, then no written plan is required.

The activities during the session are mainly focused on validating and correcting the knowledge captured at the semi-structured interview/s. The session should also be used to capture new knowledge. Hence the activities are similar to those used at the semi-structured interview (see Step 16). Therefore, you need to keep the expert focused on the project scope and help him/her recall and explain the relevant knowledge. The concept tree is an excellent way of doing this.

The key activities involve walking the expert through the tree and prompting him/her to:

- Check all terminology and rename concepts if necessary;
- Check that class membership is correct and move nodes if necessary;
- Add new concepts to the tree (classes and leaf nodes);
- Identify unnecessary concepts or synonyms that can be deleted (unless you have a 'synonyms' or 'glossary' class);
- Describe concepts that are unclear

Although it is not necessary to audio record the whole session (as most of the knowledge should be captured on the tree), it is a good idea to have recording equipment available. This is because the tree can often prompt the expert to talk about areas that were missed at the semi-structured interview, and these should be captured as an audio recording for later analysis.

Note, the use of a tree to elicit knowledge from an expert is called 'Laddering.'

4.7.5 Problems and Solutions

4.7.5.1 The Expert Gets Confused and Incorrect Relationships or Irrelevant Concepts are Added

Knowledge engineers will sometimes send the concept tree to the expert and ask for comments and additions to be made. This is dangerous. Part of the process of KA is to jointly build k-models with an expert. The partnership is very useful, since the expert knows the domain and the knowledge engineer knows how to represent the knowledge. Hence, you need to use your skills as a knowledge engineer to guide the expert to review and add to the tree. Make sure that you check 'is a' relationships are being added. Experts usually have no trouble reading and editing a concept tree if all the relationships are already 'is a'. If they are not, then confusions can often occur. Ask the expert (if needed) to check that the concepts in the tree (and those added) are within the project scope. Do not just add any concepts that the expert suggests unless you really need them.

4.7.5.2 Different Experts Give Conflicting Suggestions

If you show the concept tree to more than one expert, then conflicting opinions can occur. One expert may say that a particular concept is classified as one thing and a second expert may say it is classified as another. If this happens, you should go with one view but note down (possibly in the k-page) the difference of opinion. Later in the project, you will identify all differences of opinion and discuss them with the experts (see Step 36).

4.7.5.3 Expert Does Not Check the Whole Concept Tree

A concept tree can be quite large (with hundreds of concepts) and can therefore take a long time to go through with an expert. If time is limited, some areas of the tree may get overlooked (*i.e.* not checked) and some may only get a cursory check. If you are using a dedicated KA tool, it should have in-built functionality to help you focus the expert on each part of the tree so parts do not get missed. For example, it will allow you to hide the branches that you have already checked and only expand those that are being checked. If you are using paper, then use ticks to show which nodes have been checked.

4.7.6 Checklist for Step 20

- Ensure that the expert checks the position of all concepts on the concept tree and makes any necessary changes;
- Prompt the expert to add new concepts to the concept tree to create a full taxonomy of the relevant domain knowledge;
- Document/resolve any conflicting suggestions made by multiple experts

4.8 Step 21: Take Stock and Update Plan

4.8.1 Summary

In light of the interviews with experts and what has been learnt about the domain, it is time to take a step back from the project and assess whether anything in the scope or plan needs changing. Does the focus need to be widened, narrowed or shifted? Does the plan need changing?

4.8.2 Reasons and Conditions

By this stage you have learnt far more about the domain than you knew at the scoping and planning stages. You should use this extra understanding to review the project aims and the project plan. This step should be performed on all projects, although it can be an ongoing activity rather than just a one-off at this stage.

4.8.3 Resources

The resources required for this step are: (i) Project Proposal; (ii) Project Scope; (iii) Project Schedule; (iv) Knowledge and information gathered since Step 13.

4.8.4 Activities

In light of the last few steps (especially the interview and validation sessions) you need to assess if anything has been uncovered, explained or changed that significantly affects the project. You should ask yourself these questions:

- Is the project going to achieve all that it set out to achieve?
- Will the end-product satisfy one or more genuine needs?
- Would a change of emphasis be better for the project and for the organisation?
- Does the scope feel about right?
- Does the project seem larger or smaller than envisaged?
- Does more or less knowledge have to be captured?
- Is the project plan still applicable?
- Is the project schedule achievable?
- Is it going to be harder or easier than envisaged?
- Are the experts more or less available, willing and able to pass on the required knowledge?
- Are key resources less available than required?

Answers to these questions can be met with modifications to the project proposal, the project scope and the project schedule. The changes you might consider making are outlined in Table 4.2.

Table 4.2. Possible ways of modifying the project

Changes to proposal	Changes to scope	Changes to schedule
Modify the project aims Modify the overall approach Modify the resources Change the experts to involve Change the knowledge engineer/s	De-focus the scope (look at a wider set of knowledge) Re-focus the scope (look at other knowledge areas or tasks) Focus more on specifics and details	Allow more time or less time Alter the number of interviews to conduct Alter the order of the interview sessions Involve the experts more (or less) in creating the k-base

Once you have decided and documented any changes, you should consult and inform the relevant people. This will involve the same kind of activities that you performed in the first few steps of the project. Ensure that all key people agree with the changes and that the appropriate managers allocate any additional resources.

4.8.5 Problems and Solutions

4.8.5.1 The Project Seems to be Developing into Something Quite Different from that Envisaged During the Planning Stage
As you go up the learning curve you get a better view of the project and what can be achieved. New ideas on project direction can emerge from talking to the experts. It is sometimes better to stay with the project as proposed; at other times it is better to modify the aims and scope. Talk it over with the project manager and decide what is best for the organisation.

4.8.5.2 Key People are Losing Motivation or Interest in the Project
Loss of motivation or interest in the project can be a signal that the project is not the right one for the organisation. If so, it is time to sit down with the project manager and discuss modifying, delaying or abandoning the project. If the project has to take a back seat until more pressing activities have been accomplished, then the plan should be modified to take account of this. If you, as the knowledge engineer, are not 100% behind the project, then it is unlikely to fully succeed. If you feel you need more help, then ask for this. If you feel that the experts could contribute more, then try to get this agreed. If you feel the project and methodology are too difficult for you to achieve, then talk to your manager and try to resolve the difficulties.

4.8.6 Checklist for Step 21

- Review the project aims, scope and plan;
- Propose any changes in project emphasis and/or changes to the project plan;
- Obtain agreement for any changes from key people

4.9 Step 22: Add Descriptions to K-pages

4.9.1 Summary

Each concept should have a k-page that describes what is contained in the k-base about that concept (see Section 2.3.6 on page 18). This step adds content into a descriptive field of each k-page. This content can include one or more sentences, images, tables and hyperlinks to other pages or external files.

4.9.2 Reasons and Conditions

It is very useful to have some text describing each concept. This is essential if the deliverable is a knowledge web as these descriptions will be an important part of the web pages.

4.9.3 Resources

The resources required for this step are: (i) K-modelling tool capable of k-page editing; (ii) Documents and/or transcripts and/or experts.

4.9.4 Activities

The main activity involves adding sentences into each k-page to describe the concept of interest. The obvious method is to type the sentences yourself. There are a number of other ways that can be used:

- Copy-and-paste material from existing knowledge resources (documents, databases, websites) into the k-page. Only do this if the material is relevant to your project and is clearly presented. Do not just use your k-base to duplicate material that is already available to your end-users. It can be better to write a short description and then reference or hyperlink to a more substantial description.
- Copy-and-paste text from interview transcripts into the k-page. This is a quick way of showing what the expert said. You should put quotation marks around the text and indicate who said it and when.
- Copy-and-paste text from interview transcripts and then tidy-up the content (language and structure) to make it read more clearly (and do not include quotation marks). I have often used this technique myself, especially when the expert provides clear descriptions that do not require too much tidying.
- Ask the experts to type the descriptions themselves. This can be a powerful way of involving the experts and is especially useful if they have very specialised knowledge, are good are writing, have time available to do this and are willing.

As well as sentences and paragraphs, consider including the following:

- Images, such as diagrams and pictures. This is especially important if your k-pages are to become web pages;
- K-models that you have created, such as concept maps and process maps. Alternatively, provide hyperlinks to these.

- Tables of data, numbers or text that present information in a clear and structured manner;
- Hyperlinks on key words and phrases that link to other relevant k-pages or to external files such as documents, websites and audio/video files;
- Equations, mathematical expressions and scientific notations if these are appropriate. If your k-pages are web-based, then there are special formats (similar to HTML) for encoding these, such as MathML (for mathematics) and CML (for chemistry).

4.9.5 Problems and Solutions

4.9.5.1 Text is Badly Written (e.g. Spelling Mistakes, Grammatical Errors, Poorly Chosen Language)
Some people can write well. Others find it a struggle to put words together to create a well-crafted sentence. There are three things you can do if you are having trouble. First, many tools include spell checkers and grammar checkers. If they do, then use them. If the package does not, then consider cutting-and-pasting the text to a package that does (*e.g.* a word-processor) and checking it there. Second, show what you have written to a colleague who has good language skills. The suggestions and corrections from this person should improve the material and help to improve your skills. If available, go on a course or read material that teaches you how to write in a better way. Third, try to write short, clear sentences. Think about structuring what you want to say in terms of a bulleted or numbered list of key points. If you want to describe a complex process or idea, consider a diagram or k-model instead of sentences and paragraphs.

4.9.5.2 Descriptions are Too Short
There is no ideal length for a description as it depends on the concept and your project aims. Some require only one or two sentences, particularly if your end-product is an ontology or other formal k-base. Other descriptions should be more substantial especially if your project is to deliver a knowledge web (in which many of the web pages will be the k-pages). Try to put something in the description field of each k-page even if it is just a few comments. Alternatively, ask the expert to provide descriptions or to fill-in the k-page directly. If this is not possible, and you have nothing useful to say about a concept, then consider deleting it. This is one way of spotting aspects of the domain that are not useful to include in the k-base as concepts. Note, this may not apply to some classes in the concept tree, which may not require much description but which are useful for structural and navigation purposes.

4.9.5.3 Descriptions are Too Long
In some circumstances, a large amount of text, diagrams, pictures, graphics and equations is required to describe a concept. If this stretches to pages and pages of content, then this may be too much for the end-users to read. In this case, consider splitting the content into meaningful chunks and creating some new concepts to match these chunks. You can then cut-and-paste the content so it is spread across a number of k-pages.

4.9.6 Checklist for Step 22

- Add at least one sentence to the 'Description' field of each k-page;
- Ensure the text has the correct spelling and grammar;
- Include diagrams, images and hyperlinks where appropriate

4.10 Step 23: Create Glossary

4.10.1 Summary

Add short definitions of all domain-specific terms and acronyms.

4.10.2 Reasons and Conditions

A glossary is very useful for the end-users as it shows the basic meaning of the domain-specific terminology, acronyms and jargon used by the domain experts. The glossary is also useful for you, the knowledge engineer, to help you speak the language of the expert and help you to move up the learning curve. Becoming an expert in a field often involves the need to learn the language that experts use when they communicate with each other and when they think about issues and problems.

4.10.3 Resources

The resources required for this step are: (i) Concept tree; (ii) Documents and transcripts; (iii) K-modelling tool that allows glossary descriptions to be added to the k-base.

4.10.4 Activities

First, you should select key terms and acronyms for the glossary items. Many (perhaps all) of the glossary items will be concepts already included in the concept tree. Others might be synonyms or other terms that are used to describe the concepts. Glossary items that are not already in the concept tree, can be added to it under a class called 'Other glossary items' and/or 'Synonyms'.

Write one or two sentences that clearly define the item. These can be taken from the descriptions used in the k-pages (from the previous step). Alternatively, they can be added by either: (i) Phoning the expert and asking for short definitions; (ii) Sending a list and requesting the expert to provide short definitions; (iii) Using a web-enabled tool for experts to add definitions directly.

If needed, place the glossary items in an alphabetical list or table. If you are using PCPACK, then this is done automatically when the k-base is published.

4.10.5 Problems and Solutions

4.10.5.1 Glossary Definitions are Too Long and Complex
The ideal glossary definition contains one or two sentences. If you have more, then consider placing the remainder in the description section of the k-page. If your glossary is on a web page, then you can provide a hyperlink to the k-page so that

end-users can read more. The glossary definition should be as clear as possible. If you use domain-specific terminology within it, then make sure the terms are defined in the glossary (and provide a hyperlink if possible). If you are struggling to write clear definitions, have a look at other people's glossaries for ideas on how to structure a simple, clear description.

4.10.5.2 The Glossary Contains Irrelevant Entries

Do not define terms that are obvious (*e.g.* "A car is a wheeled vehicle."). Ask yourself: "Did I understand this term before the project started?" and "Will my end-users already be familiar with this term?". If the answer is "yes" to these questions, then remove it from the glossary.

4.10.5.3 The Glossary Does Not Contain All the Important Terms

Look though the names of the concepts in your concept tree and ask yourself these questions: "Did I understand this term before the project started?" and "Will my end-users already be familiar with this term?". If the answer is "no" to these questions, then add it to the glossary. Some people create glossaries that only contain acronyms (terms made from the initial letters of words, such as KBS and MOKA). This often misses many domain-specific terms used by experts that are unfamiliar to other people. Note, at this stage the glossary will not contain all of the terms that are required by the end of the project. More terms will be added later as you acquire more knowledge from the experts. An important test of the glossary will be in Step 33 when you show some end-users what you have already achieved (*i.e.* a prototype end-product) and gain feedback. Alternatively, you can do this sooner. Show the glossary to the end-users now and ask for feedback, and/or set-up a web page that the end-users can access and provide feedback in an ongoing manner.

4.10.6 Checklist for Step 23

- Add all of the relevant domain-specific terms (jargon and acronyms) as items in the glossary;
- Add one or two sentences to describe each glossary item;
- Consider and/or consult your end-users when creating the glossary

4.11 Step 24: Create Meta-model

4.11.1 Summary

Build a meta-model for your k-base, *i.e.* a model that describes other models. It is normally presented as a concept map that shows all the main classes in the domain and the relationships between them.

4.11.2 Reasons and Conditions

A meta-model is very useful when setting up the k-base and establishing how concepts are related to one another (using relations other than '*is a*'). Some knowledge engineers also use a meta-model to help decide what properties (attributes and values) will be used to describe particular concepts.

4.11.3 Resources

The resources required for this step are: (i) Concept tree; (ii) Generic ontology, (iii) Interview transcripts; (iv) Domain documents.

4.11.4 Activities

To create the meta-model, you should do the following:

1. From the concept tree, select the main classes of concepts. These will usually include all those under the main root node that have children.
2. Add the main classes of concepts as nodes to a concept map. If there are too many to fit onto a single diagram then use more that one. It is often a good idea to split the meta-model into two, one showing the conceptual knowledge and the other the procedural knowledge. You can also use a matrix to show a meta-model. Examples are shown in Figures 4.4 and 4.5.
3. Take a pair of nodes that are associated in some way (*e.g.* Role and Software, or Events and Resources).
4. Select a relation that describes the way the pair is associated. You can use a relation that is already in the k-base, a new relation, or the inverse of an existing relation. Ensure that the relation name allows a primitive sentence to be made, *e.g.* 'Role – uses – Software', 'Events – requires – Resources'. I usually assume the names of concepts on either side will be singular (hence 'requires' not 'require' in the previous example) so do not worry if the sentences sound odd if your class names are plural.
5. Add an arrow between the nodes on the concept map and label it with the relation name. Make sure it is an arrow (not just a line) and make sure you put a label on it that is the relation name.
6. If required, show the inverse relationship in the opposite direction, *e.g.* 'Software – used by – Roles', 'Resources – required for – Events'.
7. Repeat this for all the other pairs of nodes that you wish to relate to one another in the k-base. This decision should be driven by the project scope and by the knowledge you have collected from the experts.
8. If required, show the properties of classes as either: (i) Attributes and values, shown by using a frame instead of a simple node; or (ii) Concepts and relationships, *e.g.* show that a document has a page count using 'Documents – has page count – Page Count'.

Figure 4.4 shows three examples of meta-models using concepts maps, and Figure 4.5 shows a meta-model using a matrix.

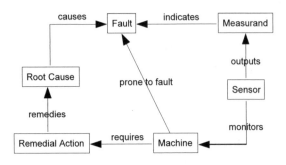

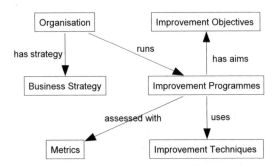

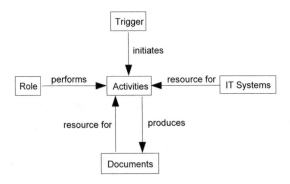

Figure 4.4. Examples of a three meta-models shown as concept maps

	Machine	Sensor	Remedial Action	Fault	Root Cause	Measurand
Machine		monitors				
Measurand		outputs				
Remedial Action	requires					
Fault	prone to fault				causes	indicates
Sensor						
Root Cause			remedies			

Figure 4.5. Example of a meta-model shown as a matrix

4.11.5 Problems and Solutions

4.11.5.1 Difficulties Selecting Which Classes of Concepts to Use Especially if There are Very Many and/or You Need Ones from Different Levels

All the relevant high-level classes in your concept tree (*i.e.* all the children of the root node) should appear on the meta-model. If you are using a generic ontology (such as GTO), then you should not include high-level classes that have no children in the concept tree. Select lower-level classes (sub-classes) of concepts if these require special relationships that are not applicable to the high-level class. If in doubt, go to a level higher in the concept tree (*i.e.* to the parent node) and ask yourself: "Does the relationship make sense here?"

4.11.5.2 Difficulties Selecting Relations to Use (Existing, New or Inverse)

The trick when selecting or creating the right relation to use between two concept classes is this: read the triple as a sentence and see if it sounds right (*i.e.* makes sense). For example, all the following sound fine: engine – part of – car; lubricant – has function – reduces friction; engine component – requires – lubricant; driver – performs – tasks. The following do <u>not</u> sound fine, so include badly named relations: engine – in – car; lubricant – does – reduces friction; engine component – required to – lubricant; driver – input – tasks. When you are doing this, do not worry if plurals or repetition make the triple sound strange, *e.g.* 'people – has role – roles' sounds wrong, but is okay, since the relationships you will be making later when you create k-models will be from individual people to individual roles, *e.g.* 'Kim Smith – has role – Finance Director'. Try to avoid very general relations such as 'associated with', 'relevant to', 'goes with' and 'linked to'. These convey little meaning and should be made more specific if possible (although they do have uses in some situations).

4.11.5.3 Difficulties in Defining Properties as Concepts or Attributes (and Using Properties of Relations)

Many knowledge engineers find it difficult to decide if attributes and values are needed. If you are inexperienced or lack confidence, then perhaps you should leave the matter of attributes and values to later in the project. Step 27 specifically deals with this and explains when and when not to use attributes and values.

4.11.6 Checklist for Step 24

- Define the high-level classes of concepts to include in the meta-model;
- Define the relationships between pairs of concept classes;
- Deal with properties of concepts

4.12 Step 25: Modify K-base Structure

4.12.1 Summary

The structure of the k-base is based on the initial template (generic ontology) used in Step 5. It is now necessary to modify the standard features of the k-base structure to allow efficient capture of more specific types of knowledge.

4.12.2 Reasons and Conditions

The templates and structures underlying the k-base require modification so that knowledge that is specific to the domain (*e.g.* relationships between certain pairs of concepts) are set-up for easy editing of the k-base and that provide automatically-generated views of k-base contents using template-driven k-models.

4.12.3 Resources

The resources required for this step are: (i) The meta-model produced on the previous step; (ii) Tool to edit the structure/templates of the k-base.

4.12.4 Activities

The activities here will depend on the k-base technology that you have and the k-base editing tool that you are using. Here are some general things that you might be able to do (some of which you may already have done):

- Modify the names of high-level classes of concepts;
- Add new high-level classes of concepts;
- Add new sub-classes to existing high-level classes of concepts;
- Add new relations;
- Alter relation properties (*e.g.* the pairs of concepts that the relation can link);
- Add new attributes and values;
- Assign attributes and values to concept classes;
- Define attributes for relations;

- Add new k-model templates (for trees, diagrams, matrices and k-pages);
- Modify existing k-model templates (for trees, diagrams, matrices and k-pages)

Some things to bear in mind when doing these changes are:

- You should set-up the k-base structure so that it matches the meta-model you defined in Step 24;
- You should make those changes that will make the entry of knowledge as quick and easy as possible;
- You should define templates for k-models so that:
    - There are a range of different ways of viewing and editing the k-base;
    - Each k-model does not try to show everything in the k-base, but provides a particular view;
    - Experts can clearly see the contents of the k-base when validating it and when providing more knowledge;
    - End-users can clearly see the contents of the k-base when assessing it and providing feedback on what else should be captured;
    - The knowledge is represented in a way that can be used when sharing the knowledge in the end-product (whether it be a website, a document or an ontology);
- Be careful when setting-up relation properties not to provide too many restrictions on the pairs of concepts that can be linked by each relation;
- Be careful when setting-up inverse relations to include all those that will be needed later;
- Be careful when setting-up the templates for trees that they do not have too many relations per tree. One relation per tree should be the norm. Only use multiple relations when you can guarantee the tree will not look confusing. Generally, it is better to have lots of small, simple trees than to have one or two large, complex trees.
- Be careful when setting-up the templates for diagrams (maps) that you get the right balance between showing enough on a diagram and showing too much on a diagram. Generally, it is better to have lots of small, simple diagrams than to have one or two large, complex diagrams.
- Edit the templates for k-pages at the highest possible level of class (as these should inherit down to all sub-types). Make sure the template includes all the information that will be stored in the k-base about that particular class of concept, *i.e.* all relationships to other concepts, all attributes, other information (*e.g.* descriptions, glossary entry) and meta-data (*e.g.* date last modified, name of author, name of expert validator). Each of these should have a separate heading. An example is shown in Figure 4.12 (page 108).
- Make use of advanced modelling techniques, such as attributes of relations, only when absolutely necessary. Keep things simple if you can.

This is one of the steps when advice from an experienced knowledge engineer will be particularly useful to you.

4.12.5 Problems and Solutions

4.12.5.1 Difficulties Altering the K-base

The problems here can stem from: you lacking the skills to edit the k-base structure; you lacking the access rights to edit the k-base structure; you lacking the support from people who do have the skills and access rights. If the latter is the case, then keep a record of the requests you make to the support staff. Use this record when contacting the project manager and support staff manager. Try to handle the situation professionally and keep those who have the power to make things happen aware of what is going (or not going) on. If the problem is with you having a lack of skills, then be proactive in finding training resources or in asking people who do have the skills to show you what to do. If you lack access rights, then make people aware of this and that this is delaying the project.

4.12.5.2 Difficulties Deciding Which K-models to Set-up

My general advice here is to ask someone who is experienced to do this for you. If your organisation is serious about knowledge acquisition, then there should be someone on the knowledge support team to help you. If there is no one to help, then there are two things you can do (depending on the technology available to you). First, do nothing. In other words, defer the creation of k-model templates to a later step when you have a clearer idea of how you want to view the knowledge. Second, do as much as you can. In other words, create as many k-model templates as you can think of. In this way, you can see what works best later, then delete those templates that are of little use. As your experience of knowledge modelling increases, you will get a better feel for which k-models work best in different project situations.

4.12.5.3 Tool Does Not Allow Required K-models

There can be situations when you wish to use a particular k-model but the software you are using does not allow this. First, check to see if the software can do what you want by contacting an experienced user or the software vendor. Second, consider an alternative model that can be created using the software. Third, consider purchasing a software tool that can do the modelling (either as a replacement for your current system, or as a plug-in). Fourth, use another existing software tool (*e.g.* a simple diagramming package) and import (or make links to) the models it creates using the k-pages in your k-base. Fifth, use paper and pencil to sketch the k-model you need. This is particularly useful when dealing with an expert at an interview either to validate the knowledge or to elicit new knowledge. Sixth, if your k-base is stored in an XML-based format then try using one of the advanced web technologies (such as XSL). This will allow special stylesheets to be written that can display the knowledge in all manner of different ways.

4.12.6 Checklist for Step 25

- Add all concept classes to the k-base;
- Add required relations, inverses and relation properties;
- Set-up a template structure for each individual k-model

4.13 Step 26: Model Relationships

4.13.1 Summary

The ways in which concepts are related to one another are added to the k-base using a number of k-models.

4.13.2 Reasons and Conditions

It is a fundamental aspect of knowledge that concepts are associated with each other. These associations are called relationships (or triples). A number of these relationships have already been captured in the concept tree using the 'is a' relation. It is now time to add other relationships. This step must be performed on all projects.

4.13.3 Resources

The resources required for this step are: (i) K-modelling tool; (ii) Interview material and/or domain documentation; (iii) Meta-model.

4.13.4 Activities

The meta-model created in Step 24 has identified the main relations to be added to the k-base. To associate a pair of concepts using these relations involves the use of one or more k-models. The three main types of k-model to use are relationship matrix, concept map and trees.

You might wish to use multiple instances of each of these types when modelling relationships. This can be done using a dedicated modelling tool, a drawing package or pen and paper. The following sections describe the use of each k-model with examples.

4.13.4.1 Relationship Matrix
The relationship matrix was described in Section 2.3.2 (page 16). Each relationship matrix is defined by the concepts you want to show as columns, the concepts you want to show as rows, and a relation. Hence, each cell in the matrix represents a single relationship (triple). An example of a relationship matrix is shown in Figure 4.6. This matrix shows a class of concepts called 'occupational group' against another called 'occupation-induced illness'. The relation used for this matrix would be something like 'susceptible to' or 'has high incidence of'.

4.13.4.2 Concept Map
Concept maps were introduced in Section 2.3.3, including an example in Figure 2.6 (page 17). Each concept map can show links representing several types of relations. You may already have used concept maps to set-up the meta-model. If so, you will have used concept maps similar to those in Figure 4.4. Another example of a concept map is shown in Figure 4.7.

	occupational group						
	clerical workers	medical workers	farm workers	factory workers	school teachers	lorry drivers	soldiers
occupation-induced illness							
musculoskeletal disorders	□		□	□		□	□
breathing and lung problems			□	□			
burns				□			□
stress	□	□			□	□	□
hearing problems			□	□			□

Figure 4.6. Example of a relationship matrix

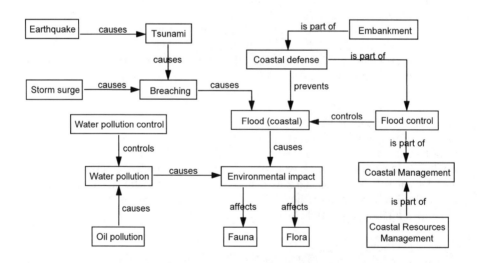

Figure 4.7. Example of a concept map

4.13.4.3 Trees
An introduction to trees was given in Section 2.3.1 (page 14). A number of different types of trees can be used to model relationships. One example is shown in Figure 4.8, which is a composition tree that uses the 'has part' or 'composed of' relation throughout. This example shows how the drive train of a mountain bike is

composed of different components and sub-components. Such a k-model is very useful when you have a triple of the form 'concept – has part - concept' in your meta-model. Thus it could be used to show the structural breakdown of such things as products, documents, organisations and projects.

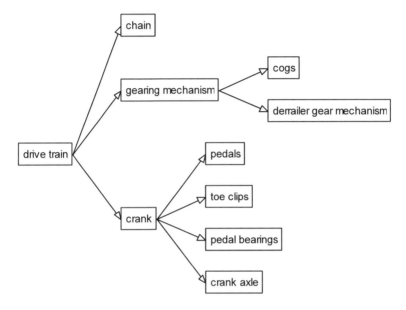

Figure 4.8. Example of a composition tree

Another type of tree that can be used at this stage is a mixed tree that shows different relations. An example of this was shown in Figure 2.3 (on page 15).

4.13.4.4 Updating the Meta-model
As you analyse and model the material from domain experts and documents you may find the need to make relationships that you had not envisaged when defining the meta-model. If this is the case (and it often is) then add the new relations and update the meta-model accordingly.

4.13.5 Problems and Solutions

4.13.5.1 Get Relationships Wrong (e.g. Confusing 'Has Part' and 'Is A')
The trick here has been mentioned before: read the triple as a sentence and see if it makes sense. When you do this, it is usually obvious when you have a good relationship and when you have a bad one. Getting relationships the wrong way around can sometimes occur. If you are using specialised software, then it can help avoid this. For instance, PCPACK allows 'relation properties' to be set-up that will restrict which classes can be linked to which other ones using a specific relation. If you are using GTO with PCPACK, then this is already built-in. A disadvantage of this may be to restrict you in what you can link to what (and this may require the

relation properties to be edited). Note, that the relation properties are a software configuration of one of the things you defined in the meta-model.

4.13.5.2 Use an Inefficient Method of Adding Relationships

There is no right or wrong way of modelling relationships, as long as your k-models do not contain errors. However, some k-models are more efficient to use than others when entering information into the k-base. The choice of which k-model to use can often be obvious. For instance, if you just want a k-model to show a single relation, then use a relationship matrix or a tree. If the relations can be shown hierarchically, then use a tree. If you wish to show a mixture of relations on a k-model, then use a mixed tree (if hierarchical) or a concept map (if non-hierarchical). Sometimes it is a case of trial-and-error to see which k-model will be the best to use in a given situation. If you are using a dedicated k-modelling tool, then this will probably be quick and easy to do.

4.13.5.3 Inadequate or Confusing K-models

If you select a k-model that suits the task at hand and you do not make any semantic mistakes (*e.g.* say that 'car – has part – elephant'), then the main problem that can occur is that the model is too large or too untidy. It is usually better to have several smaller k-models than to work with one very large one. Try to restrict the type and amount of concepts you include, especially for a relationship matrix and a concept map. Try to make your concept maps neat with no crossing links.

4.13.6 Checklist for Step 26

- Add all relationships between pairs of concepts using an appropriate k-model (*e.g.* relationship matrix or concept map);
- Add new relations, if required, and update the meta-model accordingly;
- Update k-page templates to show all the relationships that each class of concepts participates in

4.14 Step 27: Model Attributes and Values

4.14.1 Summary

Add the properties of concepts to the k-base using attributes and values.

4.14.2 Reasons and Conditions

For some classes of concepts, attributes (see page 12) and values (see page 13) are a good way of formally describing their properties. This is particularly useful for a set of concepts of the same type. The question to ask yourself is this: "Should the k-base hold detailed information on how these things differ from each other?" This step should be used selectively where necessary.

4.14.3 Resources

The resources required for this step are: (i) Project scope; (ii) Interview transcripts; (iii) Domain documentation; (iv) Tool to add attributes and values to k-base.

4.14.4 Activities

The initial activities involve: (i) Identifying sets of concepts that you may wish to describe using properties; (ii) Reviewing any attributes and values that may already have been added to the k-base; (iii) Deciding if the properties of a set of concepts are to be described using:

- Attributes and values (*e.g.* attribute=*colour* and value=*red*);
- Relations and concepts (*e.g.* relation=*has colour* and concept=*red*);
- Text sentences in the description field of a k-page (*e.g.* "The colour is always red")

To determine which of these is the most appropriate to use, follow the questioning scheme shown below.

1. Does the set of concepts contain members that are all of the same type, *e.g.* they are all vehicles, or all algorithms, or all issues? If they are not, then do <u>not</u> use attributes and values to describe them.
2. Is it important for the project that you describe the ways in which an expert thinks of the properties of the concepts? If it is not important then do <u>not</u> use attributes and values to describe them.
3. For any given attribute (*e.g.* colour), can more than one value be applied to a concept (*e.g.* the colours on the British flag are red, white and blue)? If more than one value can apply, do <u>not</u> use attributes and values in this case.
4. Could the properties of the concepts in the class be represented (and compared) using frames? To test this out, try sketching one or two frames, examples of which are shown in Table 4.3. If you cannot do this, then do <u>not</u> use attributes and values.

Table 4.3. Comparing the properties of two drinks using two frames

Coffee		Vodka	
colour	brown	colour	colourless
cost	medium cost	cost	high cost
serving temp	hot	serving temp	cold
transparency	opaque	transparency	transparent
fizziness	not fizzy	fizziness	not fizzy
amount of alcohol	no alcohol	amount of alcohol	highly alcoholic
amount of milk	no/some milk	amount of milk	no milk

If you have decided that attributes and values are a good way of describing a set of concepts, then the next activity is to add the attributes and values to the k-base. This can be achieved using a number of different k-models, such as an attribute tree (see Section 2.3.1), an attribute matrix (see Section 2.3.2) and a frame (see Section 2.3.5). This can also be achieved by identifying values from a text document such as a transcript. It is very common for experts to describe the properties of things using values but not say what the attributes are. Hence, it is often the case that you need to add in the attributes during the analysis. To do this, you should consider each value in your transcript. For each value:

1.   Add the value to the k-base;
2.   Select or add an attribute that will be associated with the value;
3.   If adding an attribute, determine what type of attribute it is: adjective, sentence, number, hypertext;
4.   If required, add other values that are associated with the attribute

These steps can be achieved using an attribute tree, an example of which is shown in Figure 4.9.

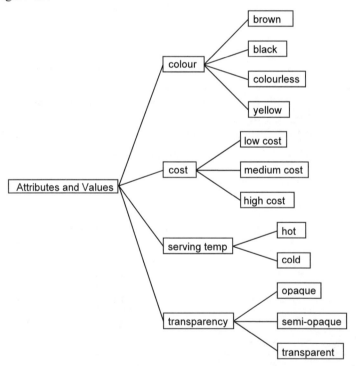

Figure 4.9. An attribute tree showing the properties of drinks

The next activity is to associate attributes and values with specific concepts. To do this, you should bear in mind the idea of inheritance. This means that attributes and values inherit down a concept tree so that members (and sub-members) of a class will have the attributes and values of the class by default. If you are unfamiliar with this idea, it is quite tricky to understand so here is an example.

- Suppose you have a concept tree showing different drinks and one branch shows different *beers*. One of the classes of beer is *lager*. A class under lager is *strong lager* and a type of this is called *Blitzenstein*.
- We want to use the attribute *amount of alcohol* to describe all drinks. So what we do is associate it with the class *drink*. This will inherit down the concept tree so that all drinks have *amount of alcohol* as an attribute.
- We now want to say that all beers are alcoholic, so we associate the value *alcoholic* with the class *beer* and this inherits down to all beers.
- We now want to say that the strong lagers are not just alcoholic but are very alcoholic, so we associate the value *very alcoholic* with the class *strong lager*, and this will inherit down to all its members, *e.g. Blitzenstein*.

In general, then, what you should do is this:

- Look at the upper levels of the concept tree and identify which attributes apply to which classes of concept;
- Associate values with upper-level concept classes if appropriate – these will inherit to sub-types as default values;
- Associate different values to lower-level concepts if appropriate, *i.e.* over-ride the default values

Once you have entered the basic attributes and values, you can use frames and/or an attribute matrix (see page 16) to view and edit the properties of concepts.

4.14.5 Problems and Solutions

4.14.5.1 Inappropriate Use of Attributes and Values or Wrong Classification of Attribute Type

One of the skills that distinguishes an experienced knowledge engineer from a less experienced one, is the ability to decide whether a certain piece of knowledge is represented as either:

- A concept;
- An attribute/value;
- A relationship;
- A piece of text in a k-page

The choice is not easy as it can depend on the context you are working in, the requirements of your end-users and the goals of the project. Take a simple word such as *colour*. The obvious decision would be to have this as an attribute with some values (red, green, *etc.*) and use it to describe the appearance of a group of concepts. For most projects, that is how you would model *colour*. For some projects, however, you may want *colour* as a concept, itself, such as in the domains of physics and optics. For other projects, you may want to use the idea of colour in a relation, hence you would represent it in the k-base using the *has colour* relation. Finally, you may not need to model colour formally (*i.e.* as an attribute, concept or relation) but just have it as a piece of descriptive text within a k-page.

4.14.5.2 Non-standard Naming of Attributes and Values

Knowledge engineers should be disciplined in the names they give to attributes and values. This helps to minimise confusions between concepts and the properties that they possess. You should name each value of an adjectival attribute as if it were an adjective. Hence, the name of a value should always be capable of putting in front of the concept name and it making grammatical sense, *e.g. fast* car, *efficient* technique, *useful* document. Since such values are equivalent to adjectives, then the names given to attributes must <u>not</u> be adjectives. Everyday examples of attributes are words such as colour, weight, and efficiency. A template sentence can be used to check the names that you have given to values and attributes. You can use the '*is a*' relation, *e.g. red* is a *colour*, *fast* is a *speed*. Alternatively, you can use the template "The *attribute* of the *concept* is *value*". Examples of this are "The *colour* of the *brake light* is *red*", "The *efficiency* of the *engine* is *inefficient*", "The *shape* of the *nose cone* is *conical*". Such sentence templates can also be used to check the names of numerical values, *e.g.* "The *weight* of the *manifold* is *360Kg*".

4.14.5.3 Poor Choice of K-model to Use to Add/Display/Edit Attributes and Values

The attributes and values of a concept can be viewed and edited using two main k-models: a frame and an attribute matrix. A frame is a very clear way of presenting the properties of a single concept. An attribute matrix is a clear way of presenting the properties of a set of concepts. Hence, the choice of which to use will generally depend on whether you wish to view/edit the properties of a single concept or a group of concepts. With a matrix, it is harder to view the properties of a single concept, but it is very quick to enter lots of values for a group of concepts. Another k-model that could be used to view the attributes and values of a concept is a concept map, *i.e.* a diagram with nodes and links. To do this, you need to represent the attribute as a relation, *e.g.* represent the attribute colour as the relation '*has colour*'. You can then have a diagram that shows 'brake light – has colour – red'. A further method uses a table or grid in which the columns and rows represent two different attributes (and the associated sets of values) and the cells contain concepts that have those values. Another way is to simply convert this table to a single dimension, so that it lists concepts with a particular value as one column and others in other columns. This is the same arrangement as would be used when card sorting (see Section 5.6.4.5 on page 128). If in doubt, try different ways of presenting the properties and see which one is best for what you are doing. This can be made much easier if you are using a dedicated modelling tool that can automatically show different k-models by using templates populated from the k-base.

4.14.6 Checklist for Step 27

- Select those classes of concepts that require attributes and values to describe their member's properties;
- Add attributes (of the right types) and values (obeying standard naming conventions) to the k-base, *e.g.* using an attribute tree;
- Assign attributes and values to the relevant concepts using an appropriate k-model, *e.g.* attribute matrix and/or frames

4.15 Step 28: Model Process Knowledge

4.15.1 Summary

Add procedural knowledge to the k-base by creating k-models that have 'tasks' as their main focus. (Note, the terms 'process', 'task' and 'activity' are used interchangeably in this book as meaning something performed to achieve a goal).

4.15.2 Reasons and Conditions

Many domains require the acquisition of procedural knowledge, *i.e.* how the experts perform certain key tasks, such as solving problems and making decisions. To do this in a clear way requires using special process-oriented k-models. This step must be performed for all projects that have included procedural knowledge in the scope. This step need not be performed if the project does not require the acquisition of knowledge from experts on how they perform activities.

4.15.3 Resources

The resources required for this step are: (i) Interview transcript/notes/sketches; (ii) Domain documentation; (iii) K-modelling tool capable of creating process maps and process trees.

4.15.4 Activities

The activities here will mainly focus on the use of three k-models: process tree, process map and k-page. You do not have to do all three, but I would recommend that you do. However, the order in which you use them is up to you. I sometimes start with process maps, then use a process tree to check and finalise the hierarchical structure. On other occasions, I do it the other way around.

In some rare situations, process maps may not be required, especially when the process is very complex with no particular order in which the tasks are performed and/or with very many iterations and loops. In this case, I suggest using a k-page to document the conditions under which each process is initiated (and perhaps also terminated). These might be in the form of rules, which you may like to represent as concepts (rather than just listing them as text in a k-page).

In general, you should do some or all of the following (although the order is up to you).

- Look through the material from the interview/s and at any other relevant documentation;
- Identify a main task that is performed by the expert/s. This may have already been done in Step 6 as part of the scoping;
- For the main task, create a process tree and add sub-tasks to it. A process tree shows the way the task decomposes into sub-tasks (see Section 2.3.1 on page 15). The depth of the tree (the number of nodes from root to leaf nodes) will be more when you require greater detail. When less detail is required, then that part of the tree can be quite shallow. An example of a process tree is shown in Figure 4.10.

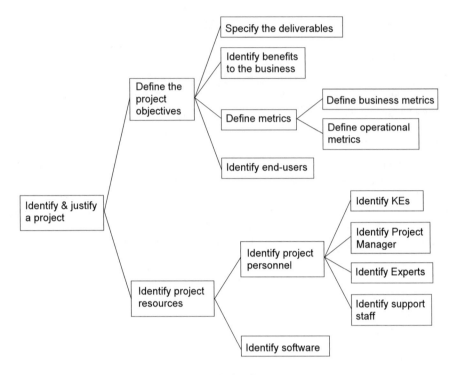

Figure 4.10. Example of a process tree

- For <u>each</u> task that has sub-tasks in the process tree, create a process map (see Section 2.3.3 on page 17).
- Each process map should show all the sub-tasks of the task as nodes. For each of these sub-tasks, consider showing some or all of this information:

  - The resources required to perform the sub-task, *i.e.* its inputs;
  - What the sub-task does, what it produces, how it changes things, *i.e.* its outputs;
  - The position of the sub-task in the process flow;
  - The conditions under which the sub-task is, or is not, performed;
  - Who and/or what performs the sub-task;
  - What triggers the sub-task to start

- An example of a process map is shown in Figure 4.11. In this diagram, the 5 rounded rectangles represent the 5 sub-tasks of the task being described. (Note, that although these are sub-tasks relative to the main task, they are all classed as 'tasks' in the concept tree). The other shapes have the following representations: rectangles represent **concepts** that act as inputs or outputs to a sub-task (such as documents); the trapezium represents a **trigger**; the diamond represents a **decision point**; the person-symbol represents a **role**. The small black down-arrow within the 'facilitate the

RCM Study' node shows that there is a lower-level process map describing this particular task (*i.e.* how to facilitate the RCM Study).

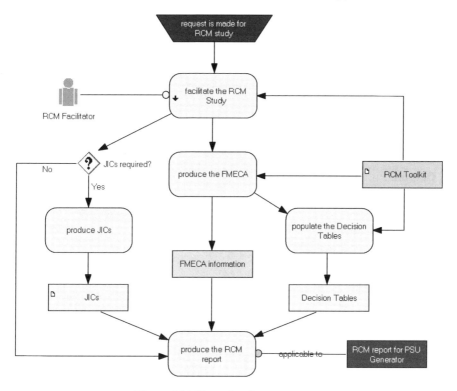

Figure 4.11. Example of a process map

- Navigate up-and-down the process hierarchy adding and re-arranging the knowledge so that the level of detail feels right. The amount shown on each process map should not overburden and confuse the viewer. The use of special modelling software can help, as it will have been designed for the purpose.
- Use k-pages to add extra information on how each task is performed, including links to process trees and process maps;
- For each low-level leaf-node task (*i.e.* one that has no sub-tasks), add instructions, guidelines, rules, equations and diagrams to its k-page to describe how it is performed. This will document the information required to perform the task (as it will not be described using a process map). Since it is a leaf node, it is usually a basic task that involves either:

  - Some very basic activities that can be described using text rather than a process map;
  - Tasks that involve looking up data or information in books, databases or graphs;

- Tasks that involve the use of algorithms or equations that can be shown and described in the k-page

An example of part of a k-page for a task is shown in Figure 4.12.

	Design the writing end
Objective	Ensure the writing end of the pen is designed to fulfil the primary functions of quality, lifetime and manufacturing cost.
Trigger	This task can start once the overall lay-up has been finalised (see define overall lay-up).
Inputs	Overall lay-up, Configuration, Nib criteria, Rollerball stick criteria.
Outputs	PDM files for nib, PDM files for head, PDM files for rollerball, plus associated documentation and diagrams.
Failure modes	Potential failures of the nib include: Nib splitting, Excessive splaying of tines, Misshaping of rollerball.
Sub-tasks	• Select the nib • Select the rollerball point stick • Design the head
Description	The nib (a.k.a. writing tip) is available in a range of styles. Each nib is created to meet the needs of specific writing styles. Therefore, it is important to first analyse the writing speed, slant, size, consistency, rhythm and pressure used when writing. Information on this is documented in Design Rules Document - WE05D. The following steps need to be performed:

Figure 4.12. Example of part of a k-page for a task

Do not worry too much if your k-models are incomplete or contain guesses. This is your first pass attempt to capture and represent the process knowledge. In the next step, you will show what you have done to the expert so that you can make the necessary corrections and prompt the expert for more knowledge. This is one of the most important uses of the k-models, *i.e.* to use with experts to elicit knowledge that is clear, correct and structured.

4.15.5 Problems and Solutions

4.15.5.1 Lack of Clarity in Naming Tasks

To be clear that a concept is a task (process, activity) and not another type of concept, then always start the name of the task with an active verb. Here are some examples of active verbs: identify, specify, decide, design, make, manufacture, investigate, inform, assess, analyse, manage, finalise, provide, perform and conduct. The rest of the name (and there should always be more than one word in a task name!) will be a phrase that includes the name of a concept. To check you have a good name for a task, put the word "they" in front of it and see if it sounds grammatically correct. For example, "*they* design the rear suspension", "*they* analyse the test results", "*they* validate the k-models" all sound good so these are good names. The following do not sound good so indicate poor names for tasks: "*they* designing", "*they* rear suspension design", "*they* model validation".

4.15.5.2 Process Maps are Too Large and/or Too Unclear

The best way to avoid having too much on each process map is to think hierarchically. In other words, try to get a good way of decomposing each task into sub-tasks, and each sub-task into sub-sub-tasks, *etc.* This is why you should construct a process tree. If you create a separate process map for each (non-leaf) node in the process tree then you will avoid having too much on each diagram. It is rarely a good idea to have more than about 10 tasks on each process map. If you have more than this, then I recommend that you try to group them into sub-tasks so that you can show them on 2 or more separate (lower-level) diagrams.

4.15.5.3 Process Maps Show Too Much or Too Little Information

There are different ways of representing process information using a diagram. For example, MOKA using a format called 'activity diagrams'. There are also international standards, such as the IDEF family of formats (*e.g.* IDEF0). The choice of which to use will be based on a number of factors: your personal preference, the methodology you are using, the process mapping tool that you use, the way your end-user wants to see the information. Let me add what I consider to be the most important factor to this list: the format that will make the acquisition and validation of the knowledge most effective and efficient. To do this, I use the form of process map shown previously in Figure 4.11. If you do not use this type of format, then you may be in danger of having process maps that lack clarity or contain too much or too little detail.

4.15.6 Checklist for Step 28

- Identify the main tasks performed by the experts (or other agents);
- Create process trees for each main task and process maps for each task that has sub-tasks;
- Add extra information to k-pages for each task, including details of how leaf-node tasks are performed

4.16 Step 29: Validate and Use K-models with Experts

4.16.1 Summary

Validation means showing k-models to experts to check the contents of the k-base are correct. This process will often involve the active editing of k-models with an expert to update the k-base and add new knowledge.

4.16.2 Reasons and Conditions

The k-base must be checked in an ongoing fashion with the experts to ensure that its contents are correct, complete, consistent and relevant to the project. The best way to do this is with k-models that allow simple, clear views of the k-base. This step must always be done (unless no experts are involved in the project).

4.16.3 Resources

The resources required for this step are: (i) K-models (on computer screen or paper); (ii) Knowledge modelling tool; (iii) Experts.

4.16.4 Activities

The validation and use of k-models can be done:

- At a face-to-face session with an expert. This is the ideal method and should always be arranged if possible. At the session, the k-models can appear: (i) On a computer screen and be edited 'live'; (ii) On paper and be edited using a pen or pencil.
- Using a virtual meeting system, such as a web-enabled conference facility that allows remote computer screens to be viewed;
- Sending documents to the expert that show the k-models and obtaining feedback by either: (i) Asking questions via a phone call; (ii) Requesting that the expert makes comments and changes on the documents; (iii) Requesting that the expert fills-in a feedback sheet.

Assuming that you are using a face-to-face session, the steps to follow are these:

1. Select the k-model or k-models to use. If you plan to use more than one k-model at a single session, then select the order of presentation.
2. Arrange the session with the expert/s. For a k-model such as a tree, map or matrix, I would allow about 30 seconds for each concept that is shown. For example, you need about 10 minutes to check a process map that has 20 nodes.
3. If paper printouts are to be used, then print-out the relevant k-models;
4. At the session, explain what will happen and what knowledge is required;
5. Explain the format of the k-model to the expert, *i.e.* what knowledge it shows and the symbols and/or arrangement that is used;
6. Walk the expert through the model. THIS IS IMPORTANT! Do <u>not</u> just put the tree, map, matrix or k-page in front of the expert and ask what

he/she thinks. The way you walk the expert through the k-model and the questions you ask will depend on the k-model. Here are some examples:

    a.   With a tree, start by explaining the top-level structure. For each branch, ask the expert to check what is there: Are all the names correct? Are any nodes in the wrong place? Use the tree to prompt for new knowledge: What important nodes are missing? Do any new nodes need adding? Should any nodes be deleted?

    b.   With process knowledge, start by showing the process tree. Guide the expert from the top level to the lower levels. Add any sub-tasks that are missing. Then move to the process maps. Start with the main, top-level tasks then move down each branch. For each map, ask the expert the same kinds of questions as for the concept tree. Use what you already have on the diagram to prompt for new knowledge. What information is needed to do this task? Is there a task that produces this information? What does this task produce? Who performs this task? Is this task always done? Under what conditions is this task not performed?

    c.   With information of attributes and values, start by showing the expert an attribute tree. You can then use any combination of attribute matrix and frames that is best to view the knowledge. Ask if any new attributes and values are needed. Using frames is usually self-evident. However, be careful when adding new attributes/values that you obey the guidelines for naming them.

    d.   With relationship information, it is often best to use a concept map or a mixed relationship tree. These are normally better than using a relationships matrix, as it is easier for the expert to read and check. The approach with a mixed tree is the same as with a concept tree, except that you must be even clearer in explaining the relations that the link lines represent. A concept map is usually self-evident, as long as you have obeyed the basic guidelines (not too large, not too complex, arrows not lines, labels on arrows showing the relation name). Once again, lead the expert through the map, and add/modify where necessary.

    e.   With k-pages, it is simply a case of letting the expert read the page and make any changes.

7.   If you are at a face-to-face session, then take an audio recording when the expert provides descriptions and explanations. Alternatively, ask the expert to directly modify the k-model. This is especially useful with k-pages as the expert can directly add/edit a description, say. Note, however, that this can be a lengthy process. Allow a few minutes for each page when planning the session.

Not that k-pages can be used successfully with a remote method rather than a face-to-face session. Unless you are using a tool that allows the expert to edit the k-pages, I suggest that you get paper printouts and give/send them to the expert for him/her to make the required changes and additions.

4.16.5 Problems and Solutions

4.16.5.1 The Best K-model for the Job is Not Being Used
The choice of the best model to use in a given situation will depend on your own preferences and the way a particular expert reacts to a k-model. Usually, the following k-models are clear and useful: any tree (except mixed trees), concept map, process map, frames and k-pages. Matrices can be successful but are not so good if they are large with few populated cells. An alternative to an attribute matrix is the use of frames and k-pages, although these only show one concept at a time. Another alternative is to use a k-model based on card sorting in which concepts in a pile represent those with a particular value. A table showing one attribute against another can also be a useful alternative. Process maps tend to be better for validation than a process tree, although the latter is good to show the expert how you have broken a main task into sub-tasks. Only use mixed trees if they are clear and well-structured (*e.g.* Figure 2.3 on page 15).

4.16.5.2 K-models are Too Large and/or Confusing
If you have obeyed the guidelines and advice given in the past few steps, then you should not have large, confusing k-models as you start this step. However, k-models can get large and confusing at a session with an expert, as new knowledge is added. Do not worry too much about this, as you will tidy up the k-model after the session. Ensure that you leave the session with a clear idea of what the expert has said about the k-models and the additions that have been made. Stick with good knowledge modelling principles, even if these stray a little (and only a little) from the language that the expert normally uses.

4.16.5.3 Expert is Inclined to Say "That's Fine" and Not Voice Worries
This problem is rare as most experts are keen to check and add to the knowledge. When it does happen, it can seem reassuring that the expert voices few changes to the k-models that you have created. However, do not be complacent. The expert's lack of comments might indicate that he/she tolerates omissions and mild inaccuracies. This may be due to: (i) The expert's nature (some people are less of a perfectionist than others); (ii) A lack of motivation from the expert for the project; (iii) A feeling of resignation that the knowledge is too difficult to represent fully and accurately. It is your job as the knowledge engineer to identify and overcome such difficulties and prompt the expert to provide the right level of validation.

4.16.6 Checklist for Step 29

- Select the k-models to use at the session, and allocate enough time to allow full use of each k-model with the expert;
- Show all parts of all k-models to the expert from which the knowledge was elicited and prompt the expert for changes and additions;
- Make the required changes and add new elements to the k-models

5

Detailed Capture and Modelling

This chapter describes the third phase of the 47-step procedure, *i.e.* Steps 30 to 39. These steps take the project from the initial modelling, through detailed knowledge acquisition and cross-validation to completion of the k-base.

5.1 Step 30: Perform Further Interviews

5.1.1 Summary

Plan and conduct further interviews with domain experts to add to the knowledge gathered previously. The further knowledge could be in new areas or in the same areas but in more detail. Analyse and model the knowledge to create a k-base that has all the basic knowledge in a highly-structured format.

5.1.2 Reasons and Conditions

For many projects a single round of semi-structured interviews, analysis and modelling is not sufficient to gather all of the required knowledge. Hence, more interviews are required. This step must be performed when more than a basic k-base is required

5.1.3 Resources

The resources required for this step are: (i) Project Scope; (ii) Project Proposal; (iii) Standard question templates; (iv) 'Who Knows What' matrix; (v) K-models; (vi) K-modelling tool; (vii) Experts.

5.1.4 Activities

The activities here will often repeat the procedures you followed on the previous few steps, but for different areas or depths of knowledge. It is usual for a number of semi-structured interviews to be performed at this stage. You should also consider supplementing (or substituting) these with techniques that directly create or use k-models. You might also use some of the more specialised techniques that can acquire deep, tacit knowledge.

The activities to perform are as follows:

1. Identify which areas of knowledge need to be acquired, using the project scope to guide you;
2. Use the 'Who Knows What' matrix to decide which experts to interview;
3. Select the type of technique to use. Information on the uses of different techniques was described in Section 2.1, and illustrated on Figure 2.1 (page 11). In brief, you can consider using one or more of these techniques:

   - Unstructured interviews when you know very little about the area;
   - Semi-structured interviews when you need to capture a rich form of explicit knowledge;
   - Use of k-models with the expert (such as trees, concept maps, process maps, decision maps, timelines and matrices) when you want to prompt and focus the expert using a visual representation. This will help elicit structured knowledge that is quick to analyse and easy to add to the k-base;
   - Use one or more of the specialised techniques that can acquire deep, tacit knowledge. These will be used later, if required, in Step 35 but can be considered for use at this stage.

4. Arrange the interview and prepare any necessary materials, such as an interview plan;
5. If you are planning a semi-structured interview, repeat the procedure you went through in Step 15;
6. Conduct the interview using the techniques described in:

   - Step 16 for a semi-structured interview;
   - Step 29 for the use of existing k-models;
   - Step 35 for the use of specialised techniques;

7. Transcribe audio-recordings using the information described in Step 17;
8. Analyse knowledge as described in Step 18, and add the new material to the k-base using k-models, as described in Steps 19, 26, 27 and 28;
9. Conduct the required validation sessions using the k-models and the process described in Step 29;
10. Loop through activities 1-9 shown above if further knowledge is required

5.1.5 Problems and Solutions

5.1.5.1 Difficulty Deciding if More Interviews are Required
The question to ask yourself is this: Does the k-base contain enough knowledge to create the final end-product or a provisional end-product that can be assessed with end-users? If you have a clear project plan and project scope then this question should be relatively easy to answer. Things become more difficult if the plan has changed and the scope has shifted. If in doubt, do two things. First talk to the

project manager about your concerns. Second, proceed directly to Steps 32, 33 and 34. The feedback from the end-users will identify which areas of knowledge need to be gathered. If this can be achieved with simple interviews, then use interviews at Step 35 rather than specialised techniques.

5.1.5.2 Experts are Now Unavailable
If the experts are unavailable then there are two alternatives. First, wait until the experts do become available and amend the project schedule accordingly. Second, re-think the project plan in light of the lack of experts and consider downsizing the end-product. Talk to the project manager about this and decide on the best course of action.

5.1.5.3 You are Unsure Which Techniques to Use with the Experts
The information presented in Section 2.1 (pages 9-12) provides some general guidelines on when to use certain techniques. The main questions to consider are as follows: Is the knowledge to be captured focused on how the expert performs specific tasks? Is the knowledge to be captured deep inside the expert's head and difficult to articulate? The answers, used in conjunction with Figure 2.1 (on page 11), should point you at the right technique. If you are unsure, then consider using a trial session in which you try out 3 or 4 different techniques and see which works the best. This is particularly useful as some experts respond better to some techniques rather than others.

5.1.6 Checklist for Step 30

- Identify the knowledge to be gathered and who to get it from;
- Prepare and conduct knowledge elicitation sessions with experts;
- Add the new knowledge to the k-base and validate it with the experts who provided it

5.2 Step 31: Finalise Main K-models

5.2.1 Summary

Complete and tidy-up the main knowledge models.

5.2.2 Reasons and Conditions

At this point in the project most of the main knowledge has been gathered and it is almost time to assess a prototype version of the end-product with a sample of end-users. Before this can be done, the k-models must be finalised and put in a presentable state for use in the end-product.

5.2.3 Resources

The resources required for this step are: (i) K-modelling tool; (ii) K-models, (iii) Project plan.

5.2.4 Activities

The activities here involve the k-models you have created as part of the elicitation and validation session with experts and the entry of knowledge into the k-base. There are a number of modifications that can be considered to ensure each k-model presents a clear picture of a certain part of the knowledge for use in the end-product. Some of the main modifications to consider are as follows:

- Make those changes required to ensure consistency across the k-models;
- Split large k-models into smaller ones (especially large trees, large concept maps and large process maps) and delete unwanted k-models;
- Tidy-up diagrams so they are easier to understand;
- Ensure no k-model can be misinterpreted because it contains ambiguities;
- Add all descriptions and other information to k-pages;
- Check spelling and grammar of all text

Some specific areas to check and modify are shown in Table 5.1

Table 5.1. Issues involved in finalising k-models

K-model	Specific Areas to Check and Modify
Concept Tree	Ensure names are correct and unambiguous, and all links are 'is a'.
Mixed trees	Do not have too many relations per tree. Use different line formats (*e.g.* colours, arrow heads) for different relations and show this information in a key or index.
Relationship Matrix	The user should not have to do a lot of scrolling to find the content.
Attribute Tree	The names of attributes and values should follow the conventions described in Step 27 (page 104).
Concept Maps	Maps should be neat and not too large. If too large, split it and provide hyperlinks from one map to another. Ensure all arrows are labelled with the relation name. Ensure all nodes are about the same size and have a consistent look.
Process Maps	Maps should be neat and not too large. If a process map is too large, break it into sections. Add these sections as new tasks, and show how to perform them using lower-level process maps.
K-pages	Check that sentences make sense and that all the spelling is correct. Break long paragraphs of text into smaller paragraphs, and use bullet points and numbered lists. Ensure there are hyperlinks to all relevant k-pages and other sources of information outside the k-base. Ensure all hyperlinks work.

5.2.5 Problems and Solutions

5.2.5.1 Difficulty Deciding Which K-models to Split or Delete
If the k-models are to be used in an end-product such as a website or a document, then it is important that they are not too large. This will make them difficult to view and for end-users to find what they need. Large trees, containing hundreds of nodes, are acceptable as long as the technology that you are using to present them can easily scroll up and down and has a 'find' function. Concept maps and process maps that contain more than 20 nodes can become confusing for users to read, especially if there are many links between nodes. A matrix should not have hundreds of columns or rows. If it does, try to segment it into several smaller matrices by breaking it into groupings of sub-concepts (*e.g.* by using the structures in the concept tree, composition tree or process tree).

5.2.5.2 Lack of Skill in Creating Good K-models, e.g. Writing Clear Descriptions in K-pages
Training courses and books (such as this one) can go a certain way to providing the skills required to create good k-models. The more experienced you become, the more skill you will gain. If you lack the experience, try to involve someone who has. Hence, show your k-models to an experienced knowledge engineer and get his/her comments.

5.2.5.3 The Tool You are Using Does Not Allow or Hinders the Clear Presentation of K-models
Some tools are excellent for creating and editing k-models, others are limited in the models they can create or require a lot of work to make them look good. If you are using tools that have not been specifically designed for knowledge engineers then you must be even more careful in what each model contains and how the knowledge is presented. You will have to be diligent in ensuring consistency across the different k-models if the tool does not help you to do this.

5.2.6 Checklist for Step 31

- Ensure all k-models are neat, clear, unambiguous and not too large;
- Ensure all k-models have the required content, especially k-pages;
- Add to existing k-models (perhaps with input from experts) to fill-in gaps

5.3 Step 32: Create Prototype End-product

5.3.1 Summary

Produce an initial version of the end-product (*i.e.* the main project deliverable) so that its contents and means of presentation can be assessed with end-users.

5.3.2 Reasons and Conditions

A prototype version of the end-product needs to be created so: (i) It can be assessed in Step 33; (ii) Problems involved in creating the end-product can be identified at this early stage.

5.3.3 Resources

The resources required for this step are: (i) K-base; (ii) K-models; (iii) Requirements for end-product (from the project proposal and project scope).

5.3.4 Activities

The focus of this step is to create something that can be presented to one or more end-users for feedback on the knowledge content. The activities will depend on the type of end-product that your project is producing: knowledge web, knowledge document or ontology. Let us take a look at the activities you would undertake for each of these.

5.3.4.1 Knowledge Web
A knowledge web is a website that presents the contents of a k-base to human end-users in a clear and navigable way. Here are some basic things to know:

- A knowledge web is similar to an ordinary website but is more structured and meaningful;
- A knowledge web is often generated automatically from a k-base using special software (*e.g.* PCPACK);
- A knowledge web is structured around a number of items of knowledge called concepts;
- Each concept has an associated page (a k-page) that describes the concept using a structured format (usually a grid);
- There might typically be around 200 k-pages in a knowledge web;
- The hyperlinks between k-pages have a meaning which is shown by the contents of the page (by the headings in the left-hand column of the grid);
- As well as pages of information, a knowledge web can show graphical representations of the way concepts link to one another (*e.g.* using trees and maps);
- A knowledge web will usually feature a number of ways of finding knowledge, such as tabs, an A-Z list, a glossary, a browser tree, a word/phrase search facility and hotlinks on diagrams.

At this stage of the project, the aim is to define enough of the website format so that an assessment of the prototype version can take place. The main aim of the assessment is on the content of the website rather than its look and feel. However, you should aim, if possible, to make the prototype as close to the end-product as you can manage at this stage. Note, you will be defining the finalised format in Step 40.

You should create a basic structure for the website, with some basic navigation features (*e.g.* A-Z list, browser). Do not spend time adding too many niceties at this stage (*e.g.* search, home page, special graphics). You just need enough to give your end-users a feel for the finished website so that they see the knowledge you have captured.

If you plan to auto-populate a web template with the contents of the k-base then you need to create a web template or select/edit an existing one. If you plan to manually create the prototype knowledge web then you should decide on the web-editing tool to use and a template to use if available.

When creating the prototype, you should make use of any standards that are available. Ideally, the software you are using will have templates that enable you to 'publish' the k-base to produce the knowledge web. If not, and you are doing things more by hand, then it is important that the k-base and knowledge web are consistent with one another. Do not edit the contents of one without updating the other!

5.3.4.2 Knowledge Document

A knowledge document is a document that details the knowledge that is required for the development of an intelligent software system, such as an expert system, a knowledge-based system or a KBE application. Hence, it should present the contents of the k-base in a clear and readable format for the software developer.

The structure of a knowledge document should be consistent with the structure of the k-base. There are two main ways to do this, which often form two main sections in the document:

- Use the structure of the **concept tree** to structure the sections that describe conceptual knowledge, *i.e.* have sections and sub-sections that mirror the structure of the concept tree (taxonomy) and describe individual concepts (except tasks);
- Use the structure of the **process tree** to structure the sections that describe procedural knowledge, *i.e.* have sections and sub-sections that mirror the process hierarchy and describe individual tasks

At the start of each section and sub-section, it is good idea to insert screenshots of the appropriate k-models (*e.g.* concept trees and process maps). This will help the reader to understand how each description of a concept links with the others. The sections that describe each concept should make use of the same kind of grid format that you used in the k-pages. You can simply cut-and-paste them from the k-base if you are creating the document manually.

If your software allows you to auto-create the knowledge document from the k-base then this will save you a lot of time. It is vital that the k-base and knowledge document are consistent with one another. Do not edit the contents of one without updating the other!

5.3.4.3 Ontology

An ontology is a specially coded version of a k-base that allows computers to read and use the knowledge. Special languages are used to encode the ontology often base on XML, *e.g.* RDF and OWL (Fensel, 2004). It is not feasible to create an ontology by hand unless it is a very small k-base or you are creating a simple trial version (which may be all you need at this stage). For the final ontology that you deliver towards the end of the project, you may need to create or obtain a software application to translate your k-base into the required format for the ontology. This can be created or obtained at this stage in the project, or left until later (Step 40).

5.3.5 Problems and Solutions

5.3.5.1 Lack of Available Time for Creating a Good Format

If you are short of time, then I suggest you simply use the k-models as your prototype deliverable. It is the content of the knowledge that is usually the most important thing to check with end-users at this stage. Hence, a well chosen set of k-models with appropriate commentary from you, will give the end-users a good appreciation of what you have captured and allow them to provide the necessary feedback ("more please in that area", "less about that"; "could you include something on...").

5.3.5.2 Lack of Adequate Standards and/or Skills in Preparing the Format

If you do not have a standard template, structure or format to use, then you will have to search for one, create one or make do without one. If you have produced a good project schedule at Step 12 then you should be fully prepared to use whatever is available or to perform the required activities. If the latter, then you should have already scheduled-in the required training. If not, seek assistance immediately, or make do with the k-models as described above.

5.3.5.3 The K-base Does Not Fit with the Standard Structure/Format/Template

If you did not foresee this problem when you created your project plan, then you must deal with it as quickly as possible. If you do not have the skills, seek advice. Make do with the k-models for the next step if you cannot create a prototype end-product. Alternatively, handcraft a small version of the prototype as a proof of concept (using a small part of the k-base) and use this in conjunction with k-models in the next step.

5.3.6 Checklist for Step 32

- Identify the requirements for the prototype end-product;
- Create or select a basic format for the prototype end-product;
- Populate the basic format with some or all of the k-base contents

5.4 Step 33: Assess Prototype End-product

5.4.1 Summary

Present a selection of end-users with the prototype end-product so they can provide feedback and ideas for the final version.

5.4.2 Reasons and Conditions

There are three main reasons for assessing a prototype version of the end-product at this stage: (i) End-users are able to identify gaps in the knowledge that can be addressed during subsequent sessions with experts; (ii) End-users are able to identify weaknesses in the ways being used to present/deliver the knowledge; (iii) The end-product gets exposure to, and buy-in from, end-users (*i.e.* they are involved and their concerns are listened to and acted upon).

5.4.3 Resources

The resources required for this step are: (i) Prototype end-product; (ii) List of end-users; (iii) The project manager.

5.4.4 Activities

For most projects, the emphasis of the assessment should be on the <u>content</u> of the end-product rather than its presentation. So the focus should be on questions such as: Will the end-product be useful? Is the knowledge at right level of detail? Is it presented in a clear way? What more knowledge could be included?

To a lesser extent, the look and structure of the end-product should be assessed. Can users find what they want? Can they navigate around? Do they like the look and feel of it? Will it do the job asked of it?

As well as end-users, it is worth asking the project manager to be involved in the assessment. Although this person is unlikely to have the direct input of the end-users, his/her comments can be useful and his/her appreciation of what is going on will add value to the finished product.

There are a number of methods that can be used to assess the prototype end-product:

- Remotely by sending the prototype (or location of it) to an end-user and asking for feedback;
- Sitting with an end-user and getting feedback by watching how he/she uses the prototype and asking for comments (at the end or as an on-going activity);
- Observing an end-user using the prototype and making formal observations;
- Observing an end-user doing a pre-determined set of activities with the prototype;
- Presenting the prototype to a group of end-users and asking for group feedback

My personal preference is to meet with end-users individually. For each end-user, I would arrange a session of about 20 minutes for a small prototype and 45-60 minutes for a large prototype. I would do these activities during the session:

- Summarise what has happened on the project to date;
- Explain what is required from the session and what will happen;

- Show the end-user the prototype and explain any information required to introduce it;
- Allow the end-user to use/view the prototype and ask him/her to describe his/her reactions to it;
- Prompt the end-user to make comments, *e.g.* "Is that good?", "Should there be more?", "What could be improved?", "What could be added?";
- Note down the comments made by the end-user, especially all of the suggested modifications and additions;
- At the end, summarise to the end-user the main points that you noted down;
- Explain what will happen next and when/how the end-user will next be involved;
- Thank the end-user

Whichever method you adopt, there should be 2 outcomes from the assessment:

- A list of requirements for the content of the end-product;
- A list of requirements for the style/format/structure/technology of the end-product

5.4.5 Problems and Solutions

5.4.5.1 Too Few/Many People are Involved in the Assessment
The number of people to involve is a balance between the amount of effort involved and the number of opinions to obtain. Usually, two or three people's opinions can get you a long way towards the feedback you require. Remember, that the more people you involve at this stage, the more people will feel an involvement in producing the end-product and the more people are likely to use it and spread the word of its usefulness. If the end-product is very large and you wish many people to see it and provide feedback, then I suggest: (i) Sending it to people individually (possibly with a short questionnaire), and (ii) Holding a group meeting which addresses a list of key issues.

5.4.5.2 The Assessment is Not Performed in a Way that Maximises the User Feedback
Whatever method you use, try to get the right balance between a relaxed situation in which end-users can use/view the prototype in different ways and the need to obtain more formal comments and suggestions. Remote methods and questionnaires can be used, but they do not allow you to observe the end-user interacting with the end-product in the way in which it is likely to be used in practice.

5.4.5.3 The Prototype End-product is Too Far from the Final Version (e.g. Lacking in Content or Functionality) to Get Useful Feedback
It can be better to show end-users your k-base in the form of k-models (trees, maps and k-pages) than present a prototype end-product that is very basic and sparsely populated. As stated previously, the main aims at this stage of the project are to check that the knowledge you have collected is going to be useful and to identify other knowledge to capture. As such, your k-models can be a good way of

providing you with this information. This can be combined with showing the end-users a limited prototype end-product, so that the end-user can provide feedback on the way the knowledge is to be presented. Thus you can get feedback on both the content (using the k-models) and the look/structure of the end-product (using a limited prototype end-product).

5.4.6 Checklist for Step 33

- Present the prototype end-product to a representative sample of end-users;
- Enable end-users to provide good feedback;
- Document the assessment results

5.5 Step 34: Take Stock and Decide on Final Tasks

5.5.1 Summary

Assess the project status using the results of the previous step and plan the final stages of the project.

5.5.2 Reasons and Conditions

It is very useful at this stage to review progress on the project and plan the remainder of the project. There are two reasons for this: (i) It is important to identify any ways in which the project is deviating from its intended aims and to put in place remedial actions; (ii) It is important to decide what actions are required to complete the project and to ensure a successful end-product.

5.5.3 Resources

The resources required for this step are: (i) Assessment results; (ii) Project Plan; (iii) Project status.

5.5.4 Activities

The first activity is to analyse, review and digest the results of the assessment exercise. Some of the questions to ask yourself are:

- What have I learnt from the assessment exercise?
- Is the project on track to achieve what it is trying to achieve?
- Will the end-product satisfy one or more genuine needs?
- Will end-users be happy with the end-product?
- Does anything major need to be changed?
- What extra/deep knowledge needs to be acquired?
- What user-interface issues need addressing?
- What technical issues need addressing?

Based on the answers to these questions, you should plan the final stages of the project. The main things to consider are as follows:

- What further knowledge acquisition sessions are required with experts?
- What validation of the k-base needs doing with experts?
- What modifications are required to k-models and k-base?
- What has to be done to resolve any technical issues?
- What has to be in place to create the end-product?
- What has to happen to release the end-product?
- How is the delivery, publicity and assessment of the end-product to be accomplished?

To plan the final stages, you and the project manager should put together a Completion Plan to ensure that the project team knows what has to be done. The Completion Plan should list all of the required actions, show who is responsible for each of them and show the date on which each one should be completed. The plan should be circulated and reviewed on a regular basis (at least once a week).

5.5.5 Problems and Solutions

5.5.5.1 The Status of the Project Against its Planned Objectives is Not Assessed in an Objective Way

Any project we work on can become part of us and it can be hard to gain an objective view of it. Hence, we can be biased towards its good points and away from its bad points. The assessment exercise in the previous step should have given you a good idea of how the end-product is going to be viewed and used by the end-users. The tendency of other people can sometimes be to accentuate the negative. Do not dismiss the comments that the end-users made. Some may be impractical to implement, but their overall impression is important. Your personal view together with the views of the end-users and of the project manager should provide a reasonably rounded view of the project.

5.5.5.2 The Initial Project Objectives Have Altered Due to Unforeseen Circumstances

The circumstances in an organisation are always changing. Some of these changes might affect your project, especially if it is a number of months since you started. If you identify changes that impact on your end-product or how you are creating it then talk it over with the project manager and decide on the best course of action. If needed, look back at the project proposal you created in Step 3 and see how it should be altered.

5.5.5.3 The Completion Plan is Too Aggressive

If you do not allow yourself enough time for all of the tasks then you run the risk of missing out key steps or rushing things. This may result in errors and mistakes. It is important that you deliver the end-product on time, but it is equally (perhaps more) important that it has the content, structure and presentation to provide maximum benefit to the organisation. Create a workable plan that does not put undue pressure on any one person and takes account of other commitments.

5.5.6 Checklist for Step 34

- Compare project status against aims;
- Identify actions resulting from the assessment of the prototype;
- Create a Completion Plan for the final stages of the project

5.6 Step 35: Capture and Model Detailed Knowledge

5.6.1 Summary

Capture detailed, deep and tacit knowledge from experts using special elicitation techniques. Analyse and model the knowledge, and add to the k-base.

5.6.2 Reasons and Conditions

By this stage most of the substantial knowledge should have been gathered using interviews and k-models. On some projects, detailed knowledge needs to be gathered from experts using special elicitation techniques. These techniques probe for knowledge that is difficult to explain during a normal interview situation.

5.6.3 Resources

The resources required for this step are: (i) Domain information, situations, cases or items; (ii) Equipment or tools for use with elicitation techniques; (iii) Audio- or video-recording equipment; (iv) Experts.

5.6.4 Activities

There are a number of techniques that can be used to elicit knowledge that is deep, detailed or tacit from experts. These techniques were introduced in Section 2.1.1.3 (on page 10).

When deciding which techniques to use, the first thing to look at is what knowledge needs to be captured. This can usually be thought of in terms of one of the following:

- Detailed **Procedural** Knowledge, *i.e.* how does the expert perform one or more complex tasks? This type of knowledge is required if your end-product is a knowledge web informing people how to perform tasks or a knowledge document informing software developers how to build an intelligent (knowledge-based) system.
- Detailed **Conceptual** Knowledge, *i.e.* what are the properties of, and relationships between, important sets of concepts? This type of knowledge is required if your end-product requires in-depth descriptions of concepts either to educate end-users, archive the information, create an intelligent system or complete an ontology.

Table 5.2 shows the techniques to be used for each of these types of knowledge.

Table 5.2. Specialised elicitation techniques used to capture deep/tacit knowledge

Procedural Knowledge	Conceptual Knowledge
Commentary (see 5.6.4.1)	Concept sorting (see 5.6.4.5)
Limited-information tasks (see 5.6.4.2)	Triadic elicitation (see 5.6.4.6)
Constrained-processing tasks (see 5.6.4.2)	Twenty questions (see 5.6.4.7)
Critical Decision Method (see 5.6.4.3)	Repertory Grid (see 5.6.4.8)
Scenarios (see 5.6.4.4)	Mapping techniques (see 5.6.4.9)

A description of each of these techniques is shown in the following sections.

5.6.4.1 Commentary

This technique involves the expert describing a task as it is performed. The basic technique here is the self-report, in which the expert provides a running commentary of his/her thought-processes as a problem is solved or a task is performed. Experimental evidence has shown that self-reports can access cognitive processes that cannot be recalled afterwards without omissions and distortion (Ericsson and Simon, 1980). The basic procedure is as follows:

1. Decide on a task that requires detailed knowledge elicitation;
2. Arrange with the expert (and/or others) to have any required resources, equipment or location available. For example, I once used the commentary technique with a military pilot. This took place in a flight simulator that the pilots used for training purposes.
3. At the session, explain to the expert what is to happen, *i.e.* the task to be performed and that he/she will provide a running commentary.
4. Ask the expert to try to say everything that he/she is thinking about, looking at and doing during the task;
5. Start the audio- or video-recording running;
6. Allow the expert to perform the task. If he/she goes quiet, ask "What are you thinking about?" or "What are you doing?"
7. If necessary, repeat the commentary for the same task (with different conditions) or a different task;
8. After the session, make a full transcription of the recording;
9. Analyse and model the knowledge using the techniques described in Step 28;
10. Validate the knowledge with the expert at a later session.

One problem with the self-report technique is that of cognitive overload, *i.e.* the mental effort required by the expert to provide the commentary can interrupt and affect his/her performance of the task. This is especially true in dynamic domains where time is critical. One way of avoiding this problem is to use an off-line reporting technique. Here the expert is shown a recording of his/her task behaviour, typically a video, and asked to provide a running commentary on what he/she was thinking and doing at the time. An advantage of this is that the video can be paused or run at slow speed to allow time for a full explanation. Variants of these reporting techniques involve a second expert commenting on the first expert's performance, or an expert commentating on the performance of a novice or trainee.

5.6.4.2 Limited-information and Constrained-processing Tasks

Limited-information and constrained-processing tasks are techniques that either limit the time or limit the information available to the expert when performing a complex task. These techniques can provide a quick and efficient way of establishing strategies and information used by the expert.

For the limited-information task, you can do the following

1. Identify a complex task to be explored;
2. Ask the expert: "If you were to perform this task, but only had three pieces of information, what would these be?"
3. After the reply, ask: "If you had three more pieces of information, what would these be?"
4. Repeat this until the expert can provide no more information

For more information on these techniques, see Hoffman (1987).

5.6.4.3 Critical Decision Method (CDM)

The Critical Decision Method (CDM) is the method of choice for a number of knowledge engineers, particular those working with the US military. It was developed by Gary Klein in the 1980s and has been used on a number of KA projects. It is based on a theory of decision-making that is the result of examining the way experts make decisions.

CDM is used to elicit knowledge of a task that an expert has performed very many times before. As such, the expert has developed a huge amount of experience for the task and will use a lot of tacit knowledge, particularly involving the subtle use of the information that is available.

CDM is an interview technique in which particular past events and incidents are examined in great detail to expose the thought-processes that the expert uses to make decisions. The focus is on non-routine incidents, the idea being that these are usually the richest source of data about the expert's capabilities (assuming less expert practitioners can handle the routine incidents). A semi-structured interview is used to examine the incident. The questions probe for: (i) The subtle cues that the expert relies upon but that can be missed by novices; (ii) The inferences and strategies that the expert used during the incident; (iii) The options that were selected and those that were rejected. For more information on CDM, see Klein (1996).

5.6.4.4 Scenarios

Scenarios are used to place the expert in specific situations in which he/she performs a task or set of tasks that are of interest to the project. There are two types of scenarios: (i) Real situations that have occurred to the expert or to other experts;

(ii) Realistic situations that could occur in the future. To set-up and use scenarios for each task of interest, the following procedure would be used:

1. Decide on a task to be explored in detail;

2. Decide on the number of scenarios to be created. A good number to use is 3 (if time permits) as it allows general patterns of problem-solving behaviour to be identified.

3. Select or create the scenarios in which the task is to be performed. These could be developed with the expert, with another expert or with someone else (*e.g.* an end-user).

4. For each scenario, write a description of it in enough detail to allow the expert to understand what is happening and have the information available to perform the task;

5. Obtain or create any supporting materials that are required to be used by the expert;

6. At the session, explain to the expert what will happen;

7. Provide the expert with the details of the scenario, such as written descriptions, photographs, domain documents, videos, maps, *etc.*

8. Provide the expert with the resources to perform the task, *e.g.* software tools.

9. Allow the expert to talk about or perform the task. You could ask the expert to provide a commentary (see earlier) or capture the expert's actions in another way (*e.g.* using video or by taking notes).

10. After the session, use the material created or recorded to analyse and model the knowledge, *e.g.* creating process maps of the actions that were taken in the scenario.

11. Use the specific scenarios to create generalised knowledge of how the task is performed and add this to the k-base, *e.g.* using process maps;

12. Validate the new knowledge with the expert at a later session.

If your end-product is a knowledge document, then the scenarios used at the elicitation session can be used later to test the intelligent system. If so, then the scenarios and the specific actions that the expert made during each scenario should be included in the knowledge document. The expert's actions will be a benchmark against which to assess the intelligent system.

5.6.4.5 Concept Sorting
Sorting techniques are an efficient method of capturing the way an expert compares and orders concepts, and can lead to the revelation of knowledge about classes, properties and priorities. The simplest form is **card sorting**. Here the expert is given a number of cards each one displaying the name of a concept. The expert is set the task of sorting the cards into piles such that the cards in each pile have something in common. Each time the cards are sorted, it will be based on an attribute and each pile will represent a value. Here is the procedure to use:

1. Decide which class of concepts you require to explore in detail, particularly their properties (attributes and values);
2. Write the name of each concept on a separate card or piece of paper;
3. At the session, explain to the expert what is to happen,
4. Ask the expert to sort the cards into piles, so that the cards in each pile are similar in some way;
5. Ask the expert to name each pile;
6. Write down (or photograph) the results of the sort (code letters or numbers on each card can help reduce the time to do this);
7. Collect the cards together and ask the expert to sort them again;
8. Repeat these steps until the expert can sort no more

For example, suppose an expert in astronomy is asked to sort cards showing the names of planets. On the first sort, the expert uses the size of the planet to distinguish them. On the second sort, the planet's surface temperature is used. On the third sort, it is the orbital velocity that is used. Each time the expert creates a pile, he/she is asked to give a name to each pile. Hence, each sort gives an attribute and the names of the piles give the values.

By sorting over and over again, many different attributes can be identified. The expert should be encouraged to sort the cards as many times as possible until the dimensions (attributes) that he/she knows about are exhausted.

Variants of this technique involve sorting objects, diagrams and photographs rather than cards. This is particularly useful in domains where simple textual descriptors are not easily available. An overview of different sorting methods can be found in Rugg and McGeorge (1997).

5.6.4.6 Triadic Elicitation (or Three Card Trick)
This technique is often used alongside the concept sorting technique. It involves prompting the expert to generate new attributes by asking the expert what is similar and different about three randomly chosen concepts, *i.e.* in what way are two of them similar and different from the other. This is a way of eliciting attributes that are not immediately and easily articulated by the expert. Here is the procedure to follow for cards (although the same procedure can be used for objects, diagrams, photographs, *etc.*):

1. Explain to the expert that you will try another technique that aims to draw out some of the deep knowledge;
2. Collect together all of the cards;
3. Shuffle them and randomly select three of the cards;
4. Place the three cards on the table so that two are next to each other, and the other is a little further way;
5. Ask the expert "In what way are these two (pointing) similar but different from this one?".
6. Write down what the expert says using an attribute, and use this to do another card sort to find the values of all the concepts for this attribute.

If the expert cannot think of an attribute to distinguish them, then no matter, just try another 3 cards;

7. Repeat these steps until the expert can think of no more differences

5.6.4.7 Twenty Questions

Rather bizarrely, the twenty-questions technique involves the expert asking questions of the knowledge engineer! This occurs because the knowledge engineer sets a puzzle in which the expert must work out the solution by asking questions.

In one variant, the questions asked by the expert provide an efficient method of exposing the properties (attributes and values) of a set of concepts in a prioritised order. If you want to use this technique, then the procedure is as follows.

1. Decide on a set of concepts that you need to explore in detail (and ensure there are very many individual types of this concept);
2. Explain to the expert what is going to happen, *i.e.* that he/she is to guess the answer to a puzzle and in so doing will use, and hence reveal, the knowledge that needs to be captured;
3. Ask the expert to try to imagine that you, the knowledge engineer, knows as much as he/she does about this set of concepts and that you have got one of them in mind;
4. Tell the expert that he/she should ask you questions to deduce the answer using the least number of questions;
5. Tell the expert that you can only answer 'yes' or 'no' to each question;
6. Explain that the best way to solve the puzzle is to ask questions that spilt the remaining concepts in half so that each answer eliminates half of the possible solutions;
7. Start the puzzle;
8. As each question is asked by the expert, write it down; a good way to do this is as a tree so that each node (question) has two branches, one for 'yes' and one for 'no'
9. When a number of questions have been asked, take the expert back to an earlier question and change the answer you gave in order to prompt the expert to ask further questions (thus exploring different areas of the question tree)
10. After the session, extract attributes and values (or new concepts) from the questions that were asked and add these to the k-base;

5.6.4.8 Repertory Grid

This technique is similar to concept sorting but allows the expert to provide ratings (scores) of each concept for an attribute rather than just placing it in one pile or another. When many ratings are made using many attributes, then statistics can be applied to find clusters and correlations. This is quite a complex technique that has been documented in many other places (*e.g.* Boose, 1988; Gaines and Shaw, 1993). The basic technique usually follows 4 main stages:

1. In the first stage, the concepts (called elements) are selected. For the technique to be successful and not take too much time to operate, the number chosen should be more than 6 and less than 15. A set of about the same number of attributes (called constructs) is also required. These should be such that the values can be rated on a continuous scale (*e.g.* from small to large, unimportant to important, expensive to very expensive). The attributes can be selected from knowledge previously elicited or generated using the triadic elicitation technique.

2. During the second stage the expert provides ratings of each concept against each attribute. A numerical scale is often used (*e.g.* $1 - 9$). For instance, if the concepts are planets in the solar system, each might be rated on its distance from the sun (1 meaning close to the sun, 9 meaning far away), and so on through the other attributes.

3. In the third stage, the ratings are applied to a statistical calculation called cluster analysis to create a visual representation of the results called a focus grid. The cluster analysis ensures that concepts with similar scores are grouped together in the focus grid. Similarly, attributes that have similar scores across the concepts are grouped together in the focus grid.

4. In the fourth stage, the knowledge engineer walks the expert through the focus grid gaining feedback and prompting for knowledge concerning the groupings and correlations. If appropriate, extra concepts or attributes are added and then rated to provide a larger and more representative grid. In this way the technique can be used to uncover hidden correlations and causal connections.

5.6.4.9 Mapping Techniques

Mapping techniques involve the use of maps, *i.e.* diagrams with nodes and links. You should be familiar with some maps from earlier in the project, such as process maps and concept maps (*e.g.* Steps 26 and 28). Here you are using maps not primarily as k-models but as a way of helping the expert to explore the relationships between concepts.

You can use any type of map or diagrammatic representation. Process maps are used when process mapping and concept maps are used when concept mapping. Other diagrams, such as decision maps and state diagrams, can also be used. An example of a state diagram is shown in Figure 5.1. In this type of diagram, the oval nodes represent states of a concept (*e.g.* the states of a telephone in this figure), and the arrows represent tasks or events that alter the state of the concept.

When using a mapping technique, I would normally use a large sheet of paper and a pencil. I would start with one or a few nodes and just explore the area with the expert. Do not worry too much if things do not start off very well, as the act of creating a map can be a trial-and-error process especially at the start. However, do not pursue the map if the expert starts to back off. If this occurs, switch to another map or another method. Remember to use arrows (and not just lines) between pairs of nodes and label each arrow with the relation it represents, *e.g.* causes, followed by, creates, connected to, has issue, transformed by, *etc.*

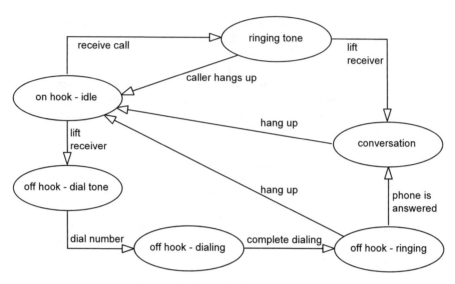

Figure 5.1. Example of a state diagram

5.6.5 Problems and Solutions

5.6.5.1 The Knowledge Engineer Lacks the Confidence or Skill to Use Special Techniques

This is a major problem that hinders the use of special elicitation techniques. Training and support are vital. If knowledge engineers feel that they fully understand the technique and have practiced it beforehand, then they will be: (i) More inclined to use it; (ii) Confident when they do use it. If knowledge engineers lack understanding and apply the technique incorrectly then it is likely to result in a poor elicitation session and reduced confidence in the techniques. If you do not have training or support available, then read what you can about these techniques and practice them with other people before trying them with the domain experts.

5.6.5.2 The Expert Feels the Technique Being Used is a Bit Silly or a Waste of Time

This is a much talked about problem although I have not experienced it myself. It can be avoided in three ways. First, fully understand when and how to use each specialised technique. Second, fully explain to the expert what will happen and why you are using the technique. Third, present the technique in a light-hearted way or in a way that warns the expert that these special techniques do not always work and are a little experimental. In this way, the expert will not be too disappointed and disheartened if the technique does not work. You and the expert should understand that acquiring deep, tacit knowledge is not easy and can often be a case of trial-and-error when finding the best way of 'unlocking' the knowledge.

5.6.5.3 The Special Technique Does Not Work

If the technique is not working then persevere, as there can often be sudden insight or change of emphasis that opens things up. However, do not persevere to a point where the expert starts to feel uncomfortable or disinterested in the process. Be careful to balance the need to try to unlock the deep knowledge and the requirement to keep the expert happy. It is always a good idea to plan two or three special techniques into a single session so that you can switch from one to another if things are not working out. Each technique is like a key. You sometimes have to try a few keys in the lock before one turns it and allows the knowledge to flood out. Experience will teach you which keys to take with you and when to switch from one to another.

5.6.6 Checklist for Step 35

- Identify detailed (perhaps tacit) knowledge that needs to be captured;
- Use one or more special techniques with the experts to capture the detailed knowledge;
- Add the detailed knowledge to the k-base

5.7 Step 36: Perform Cross-validation

5.7.1 Summary

Show k-models developed with one expert to another expert to check that the knowledge is correct, complete and relevant. Where there are different opinions and ideas, document these for the next step.

5.7.2 Reasons and Conditions

Unless your project has only one expert, then cross-validation is important. It will ensure that the knowledge is as correct and complete as possible. It is also useful to highlight differences of opinion between experts to either: (i) Resolve the differences; (ii) Reach a consensus view; (iii) Document the differences as part of the knowledge acquisition.

5.7.3 Resources

The resources required for this step are: (i) K-models; (ii) Experts; (iii) Audio- or video-recording equipment.

5.7.4 Activities

The first activity is to decide which areas of the k-base can be cross-validated and which cannot. Those areas that can be cross-validated will be the ones in which more than one person has expertise. You should have defined who has expertise in which areas during the early stages of the project (notably Step 9) and should have refined this during the project.

Having decided what can be cross-validated, you need to decide whether to cross-validate it or not. This will be dependent on the time and resources available to your project and your need to have the most accurate knowledge possible. One person's view can contain subjective ideas. Get another person's view and you obtain a degree of objectivity. If you need to establish best practices, then you need to gain the views from a number of experts. So the questions to ask of each concept and each k-model are as follows:

- Are there one or more other experts that can perform the cross-validation?
- Is there the time and resources available?
- How accurate, complete and objective does the knowledge have to be?
- Does a consensus on good/best practice need to be established?

When you have decided what needs cross-validating then you should do the following activities:

1. Select the experts to involve in cross-validating the knowledge;
2. Decide how to conduct the sessions, using the same methods as you have used for earlier validation sessions (as described in Step 29);
3. Arrange sessions with the experts if doing face-to-face cross-validation or time slots with experts if doing this remotely;
4. Plan what to do during the sessions or develop material and a schedule for the remote cross-validation;
5. If conducting a face-to-face session, then for each k-model do the following:
   a. Walk the expert through the k-model or let him/her read it (if a k-page);
   b. Audio-record or document any comments and differences of opinion that the expert makes;
   c. Make changes and/or additions to the k-model with the expert;
   d. Keep a record of which k-model you have modified, which expert was involved and what comments were made;
   e. Repeat steps a-d for each k-model
6. If conducting the cross-validation remotely, then send paper or electronic documents or have a web-enabled system to do this. Provide feedback forms for comments, or provide the expert with material to edit.

By the end of this step, you should have made a number of changes to the k-base and documented these changed for use in the next step.

5.7.5 Problems and Solutions

5.7.5.1 There are No Secondary Experts Available for the Cross-validation
If there are no other experts available then you have three options. First, carry on without any cross-validation, but make a note of this within the k-pages if this is necessary. Second, suspend the project until secondary experts become available. This should only be done if it is vital that the knowledge is cross-validated. Third, carry on the project and conduct the cross-validation when the secondary experts become available.

5.7.5.2 The Secondary Expert is Not as Experienced as the Main Expert
In some domains, there will be one main expert and a number of people with some expertise but with less experience or ability. Your judgement is required to determine whether cross-validation can be successful with these people. If in doubt, consult the project manager or line managers of the experts. It can be useful to see other experts, but treat their input as comments on the main k-models rather than making major changes to the k-base. Use the Consensus Session in the next step to get these experts together and decide on the best content for the k-base.

5.7.5.3 The Secondary Expert Agrees or Disagrees Too Much with the K-models
If the secondary expert agrees too much, then this is a problem that was previously discussed in Step 29 (on page 112). If the secondary expert disagrees too much and provides a significantly different perspective on the domain, then there are two courses of action. First, review the use of the secondary expert for cross-validation (maybe he/she does not have the expertise that you were led to believe). Second, abandon the cross-validation session (*i.e.* do not make any changes to the k-models) and move straight to a Consensus Session (see Step 37).

5.7.6 Checklist for Step 36

- Define which secondary experts to show which k-models to;
- Conduct cross-validation sessions with the secondary experts;
- Document where there is agreement and disagreement between experts

5.8 Step 37: Resolve Differences of Opinions

5.8.1 Summary

Help the experts to reach a consensus view on topics where differences of opinions have been identified.

5.8.2 Reasons and Conditions

It is essential to either resolve or document any differences of opinion between experts on the same topic. This step is only applicable if: (i) There are multiple domain experts involved in the project; (ii) Differences of opinion have been identified.

5.8.3 Resources

The resources required for this step are: (i) K-models; (ii) Interview transcripts; (iii) Information captured in Step 36.

5.8.4 Activities

The activities to perform here involve a Consensus Session in which multiple experts can discuss differences of opinion. In preparing for this session, do the following:

1. Compare k-models from different experts, **or** the changes that were made during cross-validation, **or** different opinions expressed during interviews or in documents;
2. Create a list of issues to discuss;
3. For each issue on the list, create or select a k-model that illustrates the difference of opinion;
4. Arrange the Consensus Session with the experts to discuss the differences of opinion;
5. Send an agenda, including a list of the issues for discussion to the experts

You should start the Consensus Session with a short presentation on why and how differences of opinions can occur. This should put across these ideas:

- No two people agree on everything;
- People have different experiences which lead them to have different opinions on some issues;
- Differences are a natural aspect of expertise, since no one can know absolutely everything;
- People use tacit (subconscious) knowledge when solving very complex problems, which is difficult to understand and express consciously;
- Opinions, rather than facts, are often present when information is incomplete, such as what will happen in the future

You should provide a few ground rules. For instance, say that: (i) The session should be constructive and look for consensus where possible; (ii) The attendees should acknowledge that some differences of opinion may not be resolved, but simply recorded; (iii) The attendees should be free to explain their position but also to say: "Let's agree to disagree" if consensus cannot be reached.

The discussion can now begin. You will have prepared a list of points for discussion, possibly in the form of questions. It is a good idea to have these on a computer that is projected onto a screen for all to see. Try to fill-in answers for each issue, even if it is "Some think this… and others think this…".

Go through the list of points making use of whatever k-models you have prepared or that were created during the cross-validation. For each point, ask: "Is this actually a difference of opinion?" If it is, ask: "Can it be resolved?". If it can, document the consensus. If it cannot be resolved, then document the disagreements.

You should leave the Consensus Session having captured comments for each item on the list of issues in a form that can be used to update and modify the k-models and k-base.

The final task is to update the k-models and k-base to reflect what was found during the Consensus Session.

5.8.5 Problems and Solutions

5.8.5.1 You Cannot Get Experts Together for a Consensus Session
A face-to-face meeting is the best approach to resolving differences. If this cannot be achieved, then try to arrange a web-enabled meeting or a conference call. If

these are not possible, then try to get people's written comments on the issues. If none of this can be arranged, then you should carry on with the project and resolve the differences of opinion at a later date, or leave the issues open for discussion during a later meeting or project.

5.8.5.2 The Consensus Session Does Not Run Smoothly and Agreement Cannot Be Reached
You should act as a facilitator for the meeting and try to discourage any use of emotional language and non-constructive criticisms. However, you should bear in mind that a lively and heated debate can be a good thing, and that some of the things being discussed will affect people emotionally. Be careful not to let discussions turn into arguments. Always acknowledge that people have different opinions and it is your job to capture these and look for resolution where possible. Do not think you have to solve every problem during the session. Do what you can. If an issue cannot be resolved, do not let people spend an hour discussing it. Capture what people have said and move on to the next issue.

5.8.6 Checklist for Step 37

- Prepare material for the Consensus Session;
- Conduct the Consensus Session;
- Update the k-models and document any unresolved differences of opinion

5.9 Step 38: Ensure All Knowledge is Validated

5.9.1 Summary

Ensure that every concept in the k-base has been checked for accuracy by at least one expert.

5.9.2 Reasons and Conditions

During knowledge acquisition, it can be all too easy to lose track of which concepts and k-models have been validated. It is essential that all of the knowledge included in the end-product has been checked with the experts for its accuracy, completeness and relevance.

5.9.3 Resources

The resources required for this step are: (i) K-models; (ii) K-base; (iii) Records of interview/validation sessions.

5.9.4 Activities

The activities here are quite straightforward, but essential to carry out as methodically as possible. There are two main ways to check on the validation status of the concepts:

- Create a list of all the concepts in the k-base and use it as a checklist to tick off those concepts that are complete and have been validated, and those that require further work;
- Add an attribute ('validation status') into the k-base and use k-models to associated different values to each concept. You should add this attribute to the k-page of every concept so that you have a clear record of the concept's status on its k-page.

If there are concepts that have not been validated, or have not been validated since changes where made, then remedy this with the experts. If you cannot do the validation, then enter this into the k-page, using either:

- A text comment, *e.g.* "the information shown on this page has not been fully validated";
- A value for the 'validation status' attribute, *e.g.* 'validated by 1 expert', 'validated by 2 experts', 'awaiting validation';
- An image on or across the page saying "Draft"

5.9.5 Problems and Solutions

5.9.5.1 You Have Lost Track of What Has and Has Not Been Validated
If you have not kept records of what has and has not been validated, then you will have to either revisit your experts and conduct a final validation session, or send a set of k-models to the experts and ask them for their final comments. Explain that these are the final checks of the knowledge before releasing it and that their names will be on the end-product as having validated it.

5.9.5.2 There is Too Much Validate and Not Enough Time and Resources
You should have built into your project schedule the time and resources required to perform the validation. If you did not then you will have to release your end-product with a warning that not all of the knowledge has been fully validated. If this is unacceptable, then talk to the project manager about extending the time and resources available to complete the validation.

5.9.5.3 The Expert/s are Now Unavailable for Final Validation
Once again this problem should not occur if you created a good project schedule. If unforeseen circumstances have occurred and one or more experts are no longer available then you will have to release your end-product with a warning that not all of the knowledge has been fully validated. If some has been and some has not been validated, then each concept should have a comment or an attribute associated with it in the k-base describing its validation status.

5.9.6 Checklist for Step 38

- Create a way of describing the validation status of each concept in the k-base;
- Check and enter validation status for each concept in the k-base;
- If required, communicate with experts to validate any non-validated items or to clarify any confusions

5.10 Step 39: Finalise K-models and K-base

5.10.1 Summary

Complete the k-models and k-base.

5.10.2 Reasons and Conditions

This is the last step in developing the k-base. It is important that any final changes and adjustments are made before the k-base is used to create the end-product. It is vital that the k-base created during your project contains all of the knowledge that was captured in a format that will be usable for the end-product and any other requirements.

5.10.3 Resources

The resources required for this step are: (i) K-models; (ii) K-base; (iii) KA tool, or K-base editing tool/s.

5.10.4 Activities

The activities here are highly dependent on the technology and tools that have been used to create the k-models and k-base. If you have used different tools for creating k-models (trees, maps, matrices, k-pages) then you will need to combine all of the knowledge into one central repository or link it together to create a unified k-base.

It may be that you have created many k-models but have not created a k-base. If so, then you should do this using an appropriate database or data format (such as XML). The k-base should formally describe all of the knowledge that you have captured in an integrated way. The amount of work this requires will be dependent on your particular circumstances and the requirements of the end-product.

If you have been using a dedicated KA tool that allows a range of k-models to be used to create a k-base, then this step may not be required, as all the work will have been achieved. You may just need to put the finishing touches to various k-models for use in the end-product. This is a good time to make a final check that there are no spelling and grammatical errors within k-pages, and that diagrams are neat and readable.

5.10.5 Problems and Solutions

5.10.5.1 You Have Various K-models but No K-base
The technology and tools available to you may be able to create k-models (*e.g.* using simple diagramming applications) but does not create a central store that encodes all of the knowledge in a formal, computer-readable way (*i.e.* a k-base). This may not be a problem in your particular circumstances, *i.e.* your project does not need to deliver a k-base or an end-product based on a k-base. Indeed, some people are happy to point to a group of k-models, perhaps created in different tools and formats, and say "That is my k-base". For some purposes, this may be all that is required. On the other hand, there are many situations that require a k-base to be created. If this is the case, and you do not know what to do, then seek help. Look

for either a database tool or a web-oriented schema (*e.g.* XML) that can store concepts, relationships and properties (attributes and values) in a way that is easy to create, view, edit, translate and use.

5.10.5.2 The K-base Format Does Not Allow All of the Knowledge to Be Encoded
It may be the case that limitations in the k-base format or the tool used to create it will not allow all of the knowledge you have collected to be encoded in a formal way into the k-base. For example, you may have captured rules but it is not possible to store them in a formal way. If this is the case, then you should ask yourself this question: "Based on the uses to which the k-base is to be put, does this knowledge have to be stored formally?". If the answer is yes, then you have two options. First, you should find out if you can alter the k-base format or switch to another format. Second, you should find out if an informal storage of the knowledge (*e.g.* using simple text strings) can be auto-converted to a formal scheme (*e.g.* using a parsing software application) when the end-product is created.

5.10.6 Checklist for Step 39

- Decide on the tasks required to finalise the k-models and k-base;
- Finalise the k-models and k-base;

6

Share the Stored Knowledge

This chapter describes the final phase of the 47-step procedure, *i.e.* Steps 40 to 47. These steps focus on the definition, creation, testing and release of the end-product (*i.e.* the project deliverable).

6.1 Step 40: Define and Create Format of End-product

6.1.1 Summary

There are two parts to this step. First, define how the contents of the k-base will be delivered, *i.e.* what the end-product will look like. Second, create whatever structures, formats and technological infrastructure are required for the end-product. This will build upon the work you have already performed in Step 32 and the results of the assessment in Step 33.

6.1.2 Reasons and Conditions

By this stage, all of the knowledge has been captured and placed in the k-base. The emphasis is now on the end-product, which is the main deliverable of the project. The way in which the k-base will be presented and/or used in the end-product must be defined and prepared. This is the first of three allied steps: here you will design and create the format, in Step 41 you will populate it with the contents of the k-base, and in Step 42 you will assess the prototype end-product you have created with one or more end-users.

6.1.3 Resources

The resources required for this step are: (i) K-base; (ii) User requirements; (iii) System formats/infrastructure (*e.g.* standard templates); (iv) Products and results from Steps 32 and 33.

6.1.4 Activities

The activities here are highly dependent on your particular situation: the type of end-product you are creating; the technology and tools available to you; the amount of work you performed back in Step 32; the availability of any standard templates or formats that may exist in your organisation; and the results of the assessment in Step 33.

The factors you need to consider include:

- What is the purpose of the end-product?
- Who (or what) will use the end-product?
- How will the end-product be used?
- If the end-product is to be in electronic format, what format and technology will it be based upon?
- Are there any security issues?
- What technical constraints are there?
- How will the end-users know that the end-product exists?
- How will the end-users know where to find the end-product?
- How will the end-users get access to the end-product?
- Where will the end-product be located?
- Are there any configuration management issues?
- Are there any special release procedures or quality management issues?
- How will the end-product be maintained?
- What did the assessment in Step 33 indicate about the requirements for the end-product?

Let us take a look at some of the activities that you would undertake for each of the main end-products: knowledge web, knowledge document and ontology.

6.1.4.1 Knowledge Web

During Step 32 you should have created a basic format for the knowledge web (look and basic navigation). If you did not do that at Step 32, then you should do it now (see information in Step 32 for ideas on what to consider).

Once you have a basic format, you should start to define and create the more advanced features that will increase the usability and usefulness of the web site. If you conducted the assessment in Step 33, then you will have a list of design features that you should now use to define your web site format and features.

In addition, look at the list of typical knowledge web features shown in Step 41 (on pages 145-6) and decide which ones require you to design, select or create special formats or software.

If you plan to auto-populate a web template with the contents of the k-base then you need to create a web template or select/edit an existing one. If you plan to manually create the prototype end-product then you should decide on the web editor to use and any template it may draw upon.

6.1.4.2 Knowledge Document

During Step 32 you should have created the basic structure of the document, *e.g.* the order of concepts and the use and placement of k-models. If you did not, then you should do so now (see Step 32 for information). Once you have a basic format, you should start to define and create the other features that will increase the usability and usefulness of the document. If you conducted the assessment in Step 33, then you will have a list of design features that can now be used to define the document format and features. In addition, look at the list of typical knowledge document features shown in Step 41 (on page 146) and decide which ones are required for this particular document.

6.1.4.3 Ontology

During Steps 32 and 33 you should have defined, created and tested the schema and translation software. If you did, then your main activity now is to ensure that all of the modifications and additions that resulted from the assessment in Step 33 are put in place.

If you did not define and create a prototype in Step 32 then you need to do this now. You should select or design the schema that you will use to code the contents of the k-base. You also need to define the way in which the k-base will be transformed into the ontology. You will need to obtain or create the code to do this translation. To do all this requires you to consider how the end-product will be used, *i.e.* define the way in which the end-product will integrate with and operate within the entity into which it will be embedded (*e.g.* intranet, internet, network, system).

6.1.4.4 Manual or Automatic Creation of End-product

Whichever type of end-product you are producing, you will have to decide whether to create it automatically or manually. Here are some options:

- Knowledge web: Auto-transform the k-base to a knowledge web based on a publishing template, **or** create the knowledge web manually by cutting and pasting pages from the k-base into a web-editing tool;
- Knowledge document: Auto-transform the k-base to a document, **or** cut-and-paste sections into a word processor;
- Ontology: Auto-transform the k-base to an ontology language (*e.g.* OWL), **or** manually manipulate the k-base format

6.1.5 Problems and Solutions

6.1.5.1 Difficulties Defining What the End-product Should Be (e.g. Getting Requirements from End-users)

By this stage of the project you should have had ample opportunity to obtain input from end-users. If things have gone well, you should have a set of very detailed requirements from the assessment conducted in Step 33. If not and you lack an understanding of what is required of the end-product, then you have three courses of action. First, halt the project until you can get some input. Second, carry on regardless with your best guess of what is required. Third, ask for suggestions and ideas from the project team (*e.g.* project manager and domain experts) and the knowledge support team (*e.g.* experienced knowledge engineers).

6.1.5.2 Existing Standard Structures/Templates are Unavailable or Inappropriate
If an existing structure/template is unavailable and you have tried but failed to obtain it, then you will have to create one yourself. How you do this will depend on your particular circumstances and the technology that is available. Ask around to see if anyone on the project team or knowledge support team knows of a similar structure/template you can use.

6.1.5.3 Lack of Skills in Editing/Preparing Structures/Templates
If you lack the skills then you need to obtain the skills. There may be training courses and books available. Try to find someone who has the skills to help. If not, consider using a different approach to creating the end-product. You should have foreseen all this when you put together your project plan. If you did not, then talk to the project manager about suspending the project until the required skills are in place.

6.1.6 Checklist for Step 40

- Identify the key issues and requirements for the end-product;
- Define the detailed structure and format of the end-product;
- Create the necessary structures, templates, schema and applications to enable the end-product to be created

6.2 Step 41: Create Provisional End-product

6.2.1 Summary

Produce a version of the end-product as close to the releasable version as possible for final assessment with end-users.

6.2.2 Reasons and Conditions

A provisional end-product should be tested with end-users to ensure the quality of the deliverable. Without this step, final improvements and modifications cannot be identified and made. Without this step, your end-product will not be as useable and useful as it could be.

6.2.3 Resources

The resources required for this step are: (i) K-base; (ii) Format of end-product; (iii) Delivery technology.

6.2.4 Activities

The activities here will build upon the work you performed when you created the prototype end-product in Step 32 and the work you have just achieved in Step 40. If you did not create a prototype end-product at Step 32, then read and follow the information presented in Step 32.

Assuming you did create a prototype end-product, what additional work needs to be done now? Two main things are required. First, make any modifications to the format and structure of your end-product that resulted from your assessment of the prototype end-product in Step 33. Second, make all of the final additions required for the finished product. This will depend on the type of deliverable you are creating. Here are some ideas for a knowledge web and a knowledge document.

6.2.4.1 Knowledge Web
In addition to the features described in Section 5.3.4.1 (on page 118), your knowledge web will require a good 'Home Page' so people know what is in the web site and know how to access the knowledge they need. When creating the Home Page, it will usually contain the following elements:

- Title of the knowledge web;
- Image (e.g. a logo or picture);
- Introduction: Short description of the web contents;
- Uses: Short description of who can use the web and for what purposes;
- Main areas: Hyperlinks to main pages, trees or diagrams;
- Help: Brief descriptions of how to navigate and use the web;
- People: The web author, validator and project manager;
- Dates: When the web was created, last modified and is due for review

It is a good idea to include a site map as a page in the web. This is a diagram in which the nodes are the main contents of the web site, e.g. the main k-pages and the main k-models. An example of a site map is shown in Figure 6.1

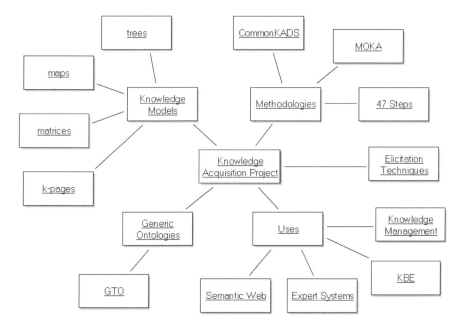

Figure 6.1. Example of a site map

Other additions to the knowledge web might include one or more of these:

- One or more browser trees to allow users to see the contents in a structured manner and find what they need;
- A categorised list of k-pages (with hyperlinks);
- A list of diagrams, such as trees and maps (with hyperlinks);
- A list of audio and video files (with hyperlinks) that allow end-users to hear or see the experts talking about the domain;
- A help page to let people know how to use the knowledge web;
- A feedback facility so that people can provide comments to the web administrator or knowledge engineer;
- An information page on the KA project describing the people who were involved, when the project took place, the projects aims and scope, *etc.*

6.2.4.2 Knowledge Document
In addition to the features described in Section 5.3.4.2 (on page 119), and those resulting from the assessment in Step 33, it is a good idea to have the following:

- A contents section or table so that people know what is in the document;
- An introductory section giving some information on the project and how the document is structured;
- A glossary of domain-specific terms and acronyms;
- An index

6.2.5 Problems and Solutions

6.2.5.1 Structure/Layout/Format of End-product is Inadequate
The way the contents of your k-base are structured (formatted, laid out, encoded) in your end-product should be driven by the requirements of your end-users. You should have had plenty of opportunity to interact with your end-users and to define the format and structure of your end-product. If this is not the case, for whatever reason, then you have the opportunity to remedy this in the next step (Step 42).

6.2.5.2 Lack of Technology or Skills for Automatic Transformation/Population
Automatically translating the k-base to create your end-product is by far the best approach if you have the technology and skills available. It is superior to a manual translation as it is much quicker and much more reliable, *i.e.* mistakes will not be made. When you created your project schedule in Step 12 you should have foreseen the need to have auto-generation of the end-product and prepared for this resource to be available when needed. If this was not the case, and auto-generation is not possible due to a lack of technology or skills, then try to find someone who can help or resort to manual means.

6.2.5.3 Manual Transformation/Population Takes an Unreasonably Long Time
If there is no available way of automatically transforming your k-base to create your end-product then you will have to do this manually, *e.g.* by copying-and-pasting from one format to another. If there is no time for this, due to lack of prior planning, or changes in circumstances, then there are three options. First, try to

find extra effort to do the translation. If this is the case, write a clear set of instructions and put safeguards in place to ensure the work is done properly. Second, delay the delivery of the end-product. Third, release a cut-down version of the end-product, with information telling the end-users when the full version will be released.

6.2.6 Checklist for Step 41

- Define the method for creating the end-product (using the k-base and format);
- Produce the provisional end-product

6.3 Step 42: Assess Provisional End-product

6.3.1 Summary

Test the provisional end-product with its target audience and/or in its target environment.

6.3.2 Reasons and Conditions

It is vital that the end-product is tested and refined before it is released.

6.3.3 Resources

The resources required for this step are: (i) Provisional end-product; (ii) End-users; (iii) Target environment.

6.3.4 Activities

You can use the same techniques that you used in Step 33, but this time the emphasis is on user-interface issues rather then the knowledge content of the end-product. Hence, you should gain feedback on the end-product's look, navigation, search, format, presentation and ease of use.

The questions to ask the end-users will be as follow:

- Is the information presented clearly?
- Is the information easy to find?
- Are there any bugs?
- Does the user get an overall feel for the structure and contents of the end-product?
- Does the user misuse or misunderstand any aspect of the end-product?
- Does the user miss out sections of knowledge?

As described in Step 33, my personal preference is to sit with each end-user and gain their comments as they use and view the end-product. The process for doing this is described in Section 5.4.4 (on pages 121-2).

At the end of the assessment you should have a list of modifications to the style, format, or structure of the end-product. If major gaps in knowledge are identified at this late stage, and experts are available, then you could consider doing some further knowledge capture. If you do, ensure the k-base and end-product both have the same additions.

6.3.5 Problems and Solutions

6.3.5.1 Difficulty Finding a Representative Sample of End-users for the Assessment
If the actual end-users are unavailable for whatever reason, and cannot be exposed to the end-product remotely for their feedback, then you should still show the end-product to one or two other people for their general comments. The project manager might provide useful input, as might an experienced knowledge engineer.

6.3.5.2 Difficulty Getting the Right Level of Feedback – Too Many Detailed
(Pernickety) Comments or Too Much "It's Fine"
All comments are welcome, however detailed they are. Whether you choose to act on all the comments is up to you in the next step. If an end-user provides little feedback, then perhaps he/she is the sort of person who is easily pleased. Alternatively, you have a great end-product on your hands! If you suspect the former, then ask the end-user to perform some real-life tasks with the end-product and observe how he/she gets on. Rather than tell you verbally what he/she thinks, you can assess the end-product by observing it being used.

6.3.6 Checklist for Step 42

- Define who will be involved and what will happen during the assessment sessions;
- Obtain feedback, comments and ideas from the end-users;
- Document the assessment results as a set of recommended modifications and improvements

6.4 Step 43: Create End-product

6.4.1 Summary

Make final modifications to the end-product (and k-base if required) and create the finalised end-product.

6.4.2 Reasons and Conditions

This step is essential as it creates the final (releasable) deliverable.

6.4.3 Resources

The resources required for this step are: (i) K-base; (ii) Formats/Templates; (iii) Delivery Technology; (iv) Assessment results, (v) Provisional end-product.

6.4.4 Activities

If you have performed Steps 32, 33, 40, 41 and 42 then the activities here should be straightforward. If you did not perform all of these steps, then you will have to perform some of the activities now.

Make the changes that were identified in the assessment of the provisional end-product. If there are too many changes to make, then prioritise them based on end-user input and the project aims. Those you do not implement should be documented so that they can be added to the end-product when time and resources become available.

Remember that the contents of the end-product and the contents of the k-base should be consistent. If you make any final changes to the end-product, then ensure these are made to the k-base.

When you have updated the end-product, and you feel it is necessary, then perform a further assessment exercise and make any final modifications.

6.4.5 Problems and Solutions

The problems and solutions will be the same as those shown in Steps 32, 33, 40, 41 and 42.

6.4.6 Checklist for Step 43

- Prioritise and select the modifications to implement;
- Implement the modifications;
- Produce the final end-product

6.5 Step 44: Release End-product

6.5.1 Summary

Perform or support the release, delivery or implementation of the end-product.

6.5.2 Reasons and Conditions

This step is required so that the end-users (people or IT systems) can access and use the end-product.

6.5.3 Resources

The resources required for this step are: (i) End-product; (ii) Deliver system.

6.5.4 Activities

The aim of these activities is to make the end-product available to end-users and/or to embed it in an IT system for use by people or software applications/services. To do this effectively will require you to deliver/implement the end-product so that is as useable and useful as possible.

Some of the specific activities that might be required are these:

- For a knowledge web: Upload the web files to an ISP/Server, Provide hyperlinks from/to an index page, portal page or the main web site;
- For a knowledge document: Print it; Distribute it; Embed it in a report; Place it in a database or on a network drive for accessibility;
- For an ontology: Deliver it; Embed it in a website/system; Make links to/from it

If required, make any minor modifications to the end-product (and k-base) to meet the implementation requirements.

You should follow any official release procedures that are necessary in your organisation. This should be relatively trouble-free if you understood what you had to do when you put your project plan together. If not, then you may have to do some extra work.

You should provide any necessary advice, training or help to the end-users and others involved with the end-product (managers, administrators, maintenance people). This could be achieved using a number of means: guide-book, presentation, training course, user manual, help pages, or an on-line demo.

6.5.5 Problems and Solutions

6.5.5.1 Delivery Mechanism is Not Compatible, Ready or Available
If you require a delivery or release procedure to be in place but it cannot be applied, then you have two courses of action. First, you must wait until it is ready and do what you can to accelerate its progress. Second, implement a pre-release of the end-product. The aim of this pre-release is to allow the end-product to be used without it being officially released or delivered. The way this is achieved will depend on your circumstances. Always remember to obey any official policies that are in place.

6.5.5.2 There is No Time to Provide Training or Help to End-users
If there is no time to do anything major (*e.g.* write a user manual or create a training course) then arrange a short informal session with one or more end-users and convey to them what you think are the main points that they need to know in order to get the best from the end-product. If you cannot do this face-to-face, note down the main points in a document and email it to the relevant people.

6.5.6 Checklist for Step 44

- Release the end-product using the relevant procedures;
- Make minor modifications to the end-product and k-base, if required;
- Provide help, training or advice to end-users;

6.6 Step 45: Publicise the End-product

6.6.1 Summary

Provide awareness to end-users and other relevant people of the end-product: what it is, where it is, how it should be used and why.

6.6.2 Reasons and Conditions

It is important that all the relevant people know about the end-product and are encouraged to use it. This is particularly important if your end-product is a knowledge web, as it is likely to have many potential users both now and in the future (perhaps some of whom you may not have thought about).

6.6.3 Resources

The resources required for this step are: (i) Descriptions of the end-product; (ii) Publicity channels/outlets.

6.6.4 Activities

Organisations normally have particular channels for publicising news and information. Find out what these are and how to make the best use of them. Ask the relevant people if you need help.

These are some of the communication channels that you might consider using:

- Presentations;
- Site-wide or corporate notices;
- Emails to specific people and/or communities;
- Messages/adverts/links on the company's intranet (for general use or for particular communities of practice/interest);
- Leaflets, flyers, or posters on notice boards;
- Articles in newsletters or magazines;

For a knowledge web that can be used by many people in an organisation, a number of more unusual publicity events might be organised, such as:

- Running a competition, such as "Hunt the Knowledge" (in which you set a number of things for people to find in your web site and offer a prize for the winner);
- Including a link to your knowledge web on a list on a web portal, *e.g.* "Top 10 Knowledge Webs", "New Knowledge Webs", "Coming Soon";
- Holding an awareness event in a 'public' area such as in a reception area or in a work's restaurant or café;

Whichever method you use, there are two things to bear in mind. First, make the publicity material short and clear. Second, do not upset people in the organisation by disobeying official procedures or accepted practices.

6.6.5 Problems and Solutions

6.6.5.1 The Publicity Material is Inadequate (Too Short/Long, Too Simple/Complex)

Writing good publicity material is a skill that many people do not possess. Seek help from anyone in the organisation who has the necessary skills. Also, test the material with one or two people to see what they think.

6.6.5.2 The Target Audience is Not Reached by the Publicity Channels

You should have a good idea who your end-users are. A simple email might be all that you need to make these people aware that your end-product is ready for use. Organising a training or awareness session (face-to-face or web-enabled) is another excellent way of reaching people.

6.6.5.3 The End-product is Publicised Once Then Forgotten About

You should put in place some permanent information to remind people of your end-product or let new people know about it. Assuming your end-product is in electronic format, an obvious way is to have a link with a short description on a web portal or index page. Also, include a link on any domain-specific web pages such as might be viewed by members of a community of practice/interest.

6.6.6 Checklist for Step 45

- Decide on the best ways to publicise the end-product;
- Prepare the publicity material;
- Deliver the publicity material

6.7 Step 46: Monitor Impact of the End-product

6.7.1 Summary

Assess the impact and usefulness of the end-product and any associated support or maintenance system.

6.7.2 Reasons and Conditions

It is important in any project to find out if the deliverables have had their intended impact. Did the project do what it set out to do? How is the organisation benefiting from the project you ran? Was it worth doing? Are there any problems with the deliverable that can be fixed and so improve its usability and usefulness? What improvements can be made to the way other KA projects are run to ensure better end-products? These issues should be examined for all released end-products.

6.7.3 Resources

The resources required for this step are: (i) End-users; (ii) Usability data, *e.g.* web statistics; (iii) Business data, *e.g.* time, resource and cost savings due to the deliverable.

6.7.4 Activities

The activities all involve an assessment of the end-product when it is being used for real in its target environment. You may have decided during the release process when and how to do this. If you did not, then you should put this in place now. Please do not release the end-product and then forget about it. You should make the assessment after a few weeks or a few months of its release, and perhaps repeat it on a regular basis after a further period of time.

You should base your assessment on the operational and business metrics that were defined in the early stages of the project and that were documented in the project proposal. If you identified business metrics that can now be measured or assessed, then you will obviously do this. This will be dependent on your situation, but will often involve obtaining figures or estimates of the amount of time, money or resources that have been saved as a result of the end-product.

There are a number of other ways of monitoring and assessing the impact of the end-product. Some methods are specific to a particular end-product and others are general. Here are some options:

- One general approach is to design and conduct a Customer Satisfaction Survey. This often involves a questionnaire that is sent to end-users or is used at a face-to-face session. This asks people how they feel about the end-product, how much they use it, how it has changed things, and how it might help the organisation in the future. It might also ask what they find good about the end-product and what could be improved.
- A second general approach is to conduct a feedback session with some of the end-users. This can be a free-form session or make use of a questionnaire or a more structured approach.
- If your end-product is a knowledge web or has resulted in a software application, then you can sit next to one or more end-users and observe their use of the web or software. To do this, you can use the methods described in Step 33 (on pages 121-2).
- If your end-product is web-based then you can obtain statistics on its use. Depending on the type of statistics your organisation operates you can get answers to a number of questions: How many hits has each page received? How did users interact with your web site? How did users locate your web site? If you need information on this approach, try searching for "web stats" or "web analytics" on an Internet search engine.

6.7.5 Problems and Solutions

6.7.5.1 Assessment Data is All Subjective or, If Objective, is Ambiguous and Interpretable in Multiple Ways
One difficulty can be disentangling the value that your end-product has contributed to an improvement when a number of other initiatives and projects may also have contributed to its occurrence. One way of doing this is to use multiple methods of assessment, such as combining objective measures of impact and usage with subjective assessments from end-users. Taking these multiple measures together will give you a more rounded picture. Another way is to talk to the people that ran

other initiatives to see how they have made assessments of the value of their projects.

6.7.5.2 Feedback is Sporadic, and Variable

Any feedback is better than none, but if the data is unclear and can be misinterpreted then it has to be treated with caution. Do not make grand claims about the success of your project if you do not have the data to back up your claims. It can be difficult to say exactly how much impact the end-product has had. A knowledge-rich product can provide benefits to an organisation that are difficult to quantify, especially as these may not become apparent for a number of years. Do the best you can, and if in doubt, get some first hand comments from end-users.

6.7.5.3 End-users are Unavailable for the Assessment Exercise

If end-users cannot be reached directly, then you will have to use other methods to assess the end-product or you will have to delay the assessment until end-users can be found. Perhaps one of the end-users is you. In that case, you should try to assess how much you have learned from the project that you ran and how this knowledge is contributing to the organisation.

6.7.6 Checklist for Step 46

- Design a way of assessing the impact of the end-product;
- Assess the impact of the end-product;
- Document the assessment results

6.8 Step 47: Document Lessons Learnt

6.8.1 Summary

Conduct a review of the project that documents the lessons learnt during the project and provides suggestions for future improvements.

6.8.2 Reasons and Conditions

It is important that lessons learnt during a particular project are captured so that they can be used to improve methodologies and support systems. This is vital, as it will prevent future projects making the same mistakes that you may have made. If knowledge acquisition is to become more efficient and deliver more business benefits, then lessons need to be learnt. This should be done on all projects.

6.8.3 Resources

The resources required for this step are: (i) Project Proposal; (ii) Project Scope and Plan; (iii) Project team including end-users; (iv) Information from Step 46 on the assessment of the released end-product.

6.8.4 Activities

There are 5 main activities:

- Decide who will run the project review;
- Plan who to involve and how to capture the lessons that were learnt;
- Capture the lessons learnt;
- Analyse the results;
- Document the results

6.8.4.1 Decide Who Will Run the Project Review

Although the knowledge engineer on the project can run the project review, it is better if someone else can do it, as this will provide a more objective viewpoint. I suggest that you ask another knowledge engineer to run the review. If the knowledge engineer on the project is the only person available, then I suggest using questionnaires or a group meeting approach in which someone else acts as the chair-person or facilitator.

6.8.4.2 Plan Who to Involve and How to Capture the Lessons Learnt

You should try to involve all of the people who took an active part in the project. The knowledge engineer, project manager and main domain experts are essential. The number of people you involve can depend on the method, or methods, you are using to collect the lessons learnt. The main methods to select from are questionnaires, one-on-one interviews (face-to-face or by phone) and a review meeting. Whichever method you use, it should probe for the information shown in Table 6.1.

Table 6.1. The information required for a project review

Perspective	Learning issue		Implications for future improvements
What went well on the project?	1.		
	2.		
	3.		
What did not go well on the project?	1.		
	2.		
	3.		

A 'Learning issue' is what happened on the project that resulted in a lesson being learnt, *e.g.* 'End-users were not involved in the project scoping stage which led to the capture of some knowledge that is not required'. An 'Implications for future improvements' is a suggestion that results from the learning issue, *e.g.* 'End-users should always be involved in the scoping stage'.

So which method should you use to capture the lessons learnt? Obviously, questionnaires are good if you have a lot of people to involve and they are distributed geographically. The downside of questionnaires is that a significant percentage of people will not respond to them. One way around this is to send out the questionnaires then conduct follow-up phone calls a few days later. You can usually capture the main lessons learnt from a single person in a 10-minute phone call.

An alternative to a phone call is a face-to-face interview, which would usually take 10-15 minutes. If doing this, I suggest you use of a questionnaire grid (similar to the one in Table 6.1) on a laptop. You can then type into the grid as the person gives you their comments. This avoids having to take an audio-recording, hence minimises the work afterwards. You should show the person what you have typed at the end of the interview to check that you have captured the main points.

A project review meeting collects people together to address the same kind of issues as shown in Table 6.1. Here are some things to consider doing at the review meeting:

- You should appoint a chair-person or facilitator to run the meeting, ideally not someone who was intimately involved in the project.
- The chair-person/facilitator should try to foster an atmosphere of constructive comments, and not criticisms, finger-pointing or blame-laying.
- The attendees should feel free to make any comments without fear or criticism.
- The comments can be captured in a number of ways: with an audio-recording for later analysis, with paper-based notes or on computer (projected onto a screen).
- The use of a questionnaire before the meeting, or at its start, is a good way for people to provide individual input that can then be discussed around the table.

6.8.4.3 Capture the Lessons Learnt

You will capture the lessons learnt using questionnaires, one-on-one interviews or a group meeting as described in the previous section. Try not to get a list of problems without recommended solutions, nor get a list of recommended solutions without the problems that they address. Ensure also that you look at what was good on the project and how this could be repeated or built upon for future projects.

6.8.4.4 Analyse the Results

My personal preference is to run the project review like a mini KA project. In this way, capturing the lessons learnt is equivalent to performing the semi-structured interviews on a proper KA project. Now you need to analyse the results. If you have used a grid like the one shown in Table 6.1, then you already have all the concepts (these are the entries in the individual cells of the grid). Using these you can create a concept tree and sort the concepts into groups. This will allow you to pull out common themes and see where people agree and disagree.

The concept tree is likely to have three main classes: learning issues, suggested improvements and areas. The latter will show the areas that each learning issue and suggested improvement relates to, *e.g.* 'planning', 'software', 'support', *etc.*

After creating the concept tree, you can link individual concepts using the following triples: 'learning issue – implies - suggested improvement', 'area - has learning issue – learning issue', 'area - has suggested improvement – suggested improvement'. You might also want to include the people who were involved and link each one to the learning issue and suggested improvement that he/she contributed.

6.8.4.5 Document the Results
You should document three things. First, the method that you employed (what you did and how you did it). Second, the key findings, *i.e.* the main learning issues and corresponding suggested improvements. Third, a list of all the learning issues and suggested improvements that were made (categorised into subject areas, such as 'planning' or 'software'). When completed, the document should be circulated to the people that took part and to the knowledge support team.

6.8.5 Problems and Solutions

6.8.5.1 There is an Over- or Under-emphasis on the Weak Areas of the Project
It is important to get the right balance during the project review between 'what went right' and 'what went wrong'. Too much emphasis on the positive, and you will miss some of the real learning points (which mainly come from the negatives). Too much emphasis on the negative, and people may leave the project thinking it was a disaster. Important lessons can be learnt from what went right, so that future projects can repeat the same things, if applicable.

6.8.5.2 Suggestions are Too Project-specific and Not Generally Applicable
A danger of using the results of a project review is to over-generalise their applicability, *i.e.* to think that all of the suggestions are applicable to all future projects. This is something you should not worry too much about during the review as it is for the knowledge programme manager and knowledge support team to decide what to do with the results. Do not try to filter what you suggest and second guess how your suggestions will be used. Instead, focus on providing the reasons for suggestions and the context in which they are being made. That it why it is important to capture the learning issues so that these can provide the rationale and context. If you only provide suggestions for future improvements then the rationale and context can be lost or misinterpreted.

6.8.5.3 Suggestions are Unworkable within the Current Set-up
A project review is more of a brainstorming activity than a planning exercise for future improvements. The people involved should be free to make any comments and suggestions whether they are practical or impractical (within limits!). It is very unlikely that all of the suggestions will be impractical. There will be some that can be implemented immediately with very little time or effort. These 'easy wins' can sometimes be just as effective as major changes. The knowledge support team will consider the other suggestions in light of the resources and strategies that are in place and decide which suggestions to pursue.

6.8.6 Checklist for Step 47

- Assess how well the project succeeded in meeting its objectives;
- Identify specific improvements, *e.g.* to the methodologies/procedures, to the training/guidance material, to the support systems, *etc.*;
- Send the suggestions for improvements to the knowledge support team

7

Implementation Issues

In this chapter I will take a look at the bigger picture of knowledge acquisition. The material is aimed primarily at experienced knowledge engineers and managers who need to set-up and run a knowledge programme within an organisation. It provides advice on a number of topics, such as developing and managing a knowledge programme, software tools for knowledge acquisition, training and supporting knowledge engineers, and defining specific methodologies and procedures. It ends with some closing remarks about the whole book.

7.1 Developing and Managing a Knowledge Programme

7.1.1 The Place of Knowledge Acquisition in an Organisation

An organisation can use knowledge acquisition (KA) in a number of different ways. Some of these will have more impact and funding than others. Here are four levels that KA can operate:

- **Level 1**. KA is used on a small number of projects that require the capture and use of knowledge;
- **Level 2**. KA is used on a wide basis to support the transfer of knowledge and raising of skill levels;
- **Level 3**. KA is used as part of a major knowledge initiative to provide significant business benefit by harnessing the information and expertise within the organisation;
- **Level 4**. KA is used to create and maintain a knowledge-enabled organisation that has at its heart a 'knowledge core'. This knowledge core fully supports the culture, processes and working practices in the organisation.

Level 1 can probably be run using a methodology like the 47-step procedure and little else in terms of methods, management, infrastructure and resources. Levels 2-4, however, require other things to be in place, notably a 'knowledge programme'. This is a defined infrastructure that provides the basis for knowledge engineers to perform successful KA projects. The knowledge programme requires a knowledge support team to run it. This is a group of people whose roles include training, supporting and managing the knowledge engineers.

7.1.2 How to Develop a Knowledge Programme

How does an organisation create a knowledge programme from scratch? In my experience, four phases of work are required. These are described in brief below.

- **Phase 1: Define Vision.** Identify a Knowledge Steering Group, *i.e.* a team of people that will make decisions and have the influence to make things happen. Define the remit of the Knowledge Steering Group, *e.g.* when it will meet and how it will communicate. Educate and inspire the Knowledge Steering Group, using presentations, discussions, workshops and a short demo project/initiative if required. Develop a vision for the programme: Where do we want to be in 1-2 years time? How do we get there? Identify the resources required to realise the vision. Identify the methods and tools to use. Conduct a knowledge audit: Where is the knowledge in the organisation? Identify a trial project to implement the vision on a small-scale. Obtain backing from senior management for the knowledge programme and trial project.

- **Phase 2: Run Trial Project.** Plan the trial project: Identify a community and domain; Define the end-users; Define the end-product/s, *e.g.* website, ontology or intelligent system; Scope what specific areas of knowledge will be involved; Define the methods and tools to use; Define a schedule; Disseminate the plan and allocate resources. Perform the trial project. For example: Learn the basics; Plan and perform interviews with experts; Analyse, model and validate the knowledge; If required, capture deep/tacit knowledge; Finalise the knowledge base; Use the knowledge base to create one or more end-products; Assess, refine, release and publicise the end-products.

- **Phase 3: Plan Rollout.** Assess the trial project: What lessons were learned? What is the impact of the end-products? How should things have been done differently? Refine the vision: Redefine and flesh-out the initial vision in light of the trial project; Establish a more accurate assessment of the required resources. Develop a plan for the rollout: Where do we go next? How should we do it? Define a schedule and resources.

- **Phase 4: Perform Rollout.** Do what was done on the trial project but bigger and better. Join up the dots: Create an interconnected 'web' of projects and knowledge bases. Keep monitoring and improving: What do those involved think? What do the metrics say about the benefits of the programme to the organisation? What can be extended or re-used into other areas of the organisation? What other initiatives are possible?

7.1.3 Managing a Knowledge Programme

Once a knowledge programme is up and running, what are the responsibilities of the knowledge support team? Here are the main ones that directly impact the efficiency and effectiveness of KA projects:

- Define one or more KA methodologies or procedures;
- Define a way of classifying projects (a 'project classification scheme');
- If you have multiple methodologies or procedures, or variants of these, then define a way of matching project types to methodologies;
- Help the knowledge engineer to select/create a procedure for a particular project;
- Define a quality procedure to ensure projects deliver and do not stray or stagnate (for example have 'review gates' at various points in a project);
- Provide software tools that allow knowledge engineers to achieve a consistently high standard of work, be reasonably autonomous and produce high quality end-products as efficiently as possible
- Provide training courses to teach people how to be knowledge engineers;
- Provide continuing support to assist knowledge engineers during their projects;
- Create awareness material to educate the organisation in the uses, benefits and techniques of KA;
- Ensure the results and suggestions from project reviews (see Step 47) are examined and lead to improvements

In the next sections, I will go into more detail about methodologies and procedures, software tools, and training and supporting knowledge engineers.

7.2 Methodologies and Procedures

Before discussing methodologies and procedures, let me clarify the difference between them. A procedure defines all of the steps in detail and in the order in which they are normally performed during a project. A methodology gives the elements required to carry out a project, but does not provide detailed advice for every step along the way, and indeed may not be clear about all of the steps that are required and the order of them.

Most KA methodologies and procedures provide a way of conducting a project from start to finish. A typical methodology/procedure defines:

- A number of project phases;
- A number of high-level knowledge objects (concepts, tasks, relations);
- A number of knowledge model formats;
- An approach or strategy for acquiring, modelling and using the knowledge

Some also provide: (i) Checklists or worksheets to follow and fill-in; (ii) Detailed activities and/or steps to follow; (iii) Sets of criteria to use at key points in a project to provide review gates.

The main KA methodologies and procedures that are in the public domain are CommonKADS, MOKA and the 47-step procedure described in this book. The main difference between these approaches is in the type of projects they aim to support (as was described in Section 3.11.4 on page 53). Let me give you some more information about CommonKADS and MOKA and then say how you might create a methodology or procedure for your own situation.

7.2.1 CommonKADS

CommonKADS is a methodology for the development of AI systems, such as expert systems, knowledge based systems and decision support systems. These systems perform tasks normally performed by a human expert, *e.g.* diagnosing an illness, designing a lift, or planning a military mission.

CommonKADS started life as the KADS methodology in the 1980s and developed into CommonKADS in 1990s as part of various European research projects. It provides a model-driven approach that:

- Identifies knowledge-oriented problems and opportunities;
- Assesses the feasibility of a knowledge system;
- Models the organisation, the system and the knowledge;
- Uses a library of task templates (*i.e.* general ways of solving knowledge-intensive problems)

CommonKADS is described in detail in the book 'Knowledge Engineering and Management' (Schreiber *et al.*, 2000).

7.2.2 MOKA

MOKA stands for Methodology and software tools Oriented to KBE Applications. It is a methodology for the development of software applications that are used in computer aided design (CAD) systems. These applications are known as knowledge based engineering (KBE) applications. MOKA provides a structured software-assisted approach to:

- Identify relevant knowledge from documents and experts (who are usually design engineers);
- Create a number of specific knowledge models (*e.g.* diagrams);
- Publish and disseminate knowledge using a collection of k-pages (called 'ICARE forms') using a web site (called a 'knowledge book');
- Develop KBE applications

It has two stages: an informal modelling stage that captures knowledge and creates a knowledge book, then a formal stage that re-models the knowledge to create a set of formal diagrams that are used to design a KBE application.

MOKA is described in detail in the book 'Managing Engineering Knowledge' (Stokes, 2001).

7.2.3 Creating and Combining Methodologies

As mentioned earlier in this book, the 47-step procedure can be used in a number of ways: (i) You can use it as the procedure to follow when performing a KA project; (ii) You can incorporate aspects of it into another approach that you are using, such as CommonKADS or MOKA; (iii) You can use it as material for developing your own procedure.

If you wish to do the first of these, then I hope the information in Chapters 3-6 will give you most of what you need. If you wish to do the second or third, then some ideas were described in Section 3.11 (on pages 52-3). Here are some more ideas on how you might modify the 47-step procedure for your own requirements:

- If you have a set of generic models or ontologies (such as the problem-solving models of CommonKADS), then the 47-step procedure should be modified to emphasise: (i) The selection of the appropriate generic model; (ii) The use of the generic model to drive the elicitation of knowledge from the experts;
- If the end-product of a project is to be used to develop a software program, then particular modelling formats may be required on top of the usual k-models. For example, MOKA uses a group of special diagram formats based on UML (Booch *et al.*, 2005).
- If you have no experts to involve in a project, then all of the knowledge will come from documents and/or databases. If so, then you may need to use special 'data mining' methods and tools to extract the relevant information. Alternatively, you will have to do a lot of knowledge analysis (by hand) as a replacement for the interviews with experts.
- If you want to involve the experts more, then one way is to get them to create and edit k-models directly (without the presence of a knowledge engineer). If so, you will need to use methods and support tools that provide the experts with the capabilities to do this without taking too much time from their normal activities. One such way is to provide them with simple forms to fill-in or to provide them with user-friendly software tools.
- If the knowledge support team requires control of projects, then 'review gates' can be introduced at keys stages in the 47-step procedure. These gates are meetings at which an experienced knowledge engineer will judge the status of the project against a defined set of criteria. The project can only progress to the next stage if all the criteria have been met.

7.3 Software Tools for Knowledge Acquisition

Let us now look at some of the software tools that can be used to help a knowledge engineer perform various parts of a KA project.

7.3.1 PCPACK

PCPACK is a comprehensive tool for creating a k-base and using it to produce an end-product. It is the support tool for the MOKA and CommonKADS methodologies. I should admit that I have been part of the development team of PCPACK and it is the software I use for KA projects (so excuse me if I may be biased towards it!).

The roots of PCPACK lay in research work performed in AI and Psychology. Hence, it is a tool that is fully consistent with the methods and techniques described in the 47-step procedure. PCPACK has the following main functions:

- It supports knowledge analysis using virtual marker-pens (see Figure 4.2 on page 75,which is a screenshot from PCPACK);
- It allows all of the main k-models to be used for creating and editing a k-base. It was used to create the k-models shown in this book, such as: the concept tree on page 14; the composition tree on page 99; the attribute tree on page 102; the process tree on page 106; the mixed tree on page 15; the concept maps on pages 17, 92 and 98; the process map on page 107; the site map on page 145; the relationship matrix and attribute matrix examples on pages 16, 48 and 98; and the k-page on page 108. It also allows frames to be edited and incorporated into concept maps and process maps.
- It creates a k-base in XML format that can be published automatically as a knowledge web or a knowledge document, or can be converted to an ontology format such as OWL, or output as computer code.

For more information on PCPACK, see http://www.epistemics.co.uk/

7.3.2 Protégé

Protégé is a widely used tool for the development of ontologies. It has been developed by a team at Stanford University and has the advantage of being open source and freely available as a download. It is based on a technology (Java) that allows users to add software applications to the basic code so as to enhance its functionality.

Protégé uses a form-based editor, *i.e.* the user enters information on a concept by filling in a form that can be customised for a particular project. It is limited in the k-models that it can create, although some users have developed add-ons, *e.g.* that display a concept tree. It supports frame-based or OWL-based k-bases, and a number of other formats. Some say it lacks robustness for large k-bases so it may not be suitable for large KA projects.

For more information on Protégé, see http://protege.stanford.edu/

7.3.3 Others

Other software tools that could be used on KA projects are:

- **CmapTools**. This is a tool to build concept maps. It is free to download and can be used with a special server (also free to download) so that concept maps can be shared with other people. For more information see http://cmap.ihmc.us/
- **Repertory grid tools**. If you want to use the repertory grid technique to elicit knowledge from experts, then a software tool is essential to perform the statistical analysis. A number of applications can do this, *e.g.*

REPGRID and WEBGRID, information on which can be viewed at http://repgrid.com/
- **UML tools**. IBM produces a number of commercial tools that support the Unified Modeling Language (UML) methodology. Although the purpose of these products is to aid the design of software programs, the visual style could allow it to provide support for the modelling stages of a KA project. For more information, see http://www-306.ibm.com/software/rational/

7.4 Training and Supporting Knowledge Engineers

7.4.1 Training Knowledge Engineers

In my experience, a 3-day, classroom-based training course is enough to teach people the basics of KA. The aim is to give people who have no prior exposure to KA the basic skills necessary to be a knowledge engineer. The course has the following objectives:

- To understand the uses and benefits of KA;
- To learn KA skills, *e.g.* interviewing, analysis and modelling;
- To learn how to plan and run a successful KA project;
- To learn how to the use the relevant KA methodology and software tools;
- To develop skills in applying knowledge capture techniques and KA software on a practical project;

A typical course timetable is shown in Table 7.1.

Table 7.1. Timetable for a 3-day course for new knowledge engineers

Day 1	Morning	Introduction to the course	15 mins
		Overview of Knowledge Acquisition	1.5 hours
		Knowledge Analysis and Modelling	2 hours
	Afternoon	Introduction to Software Tools	1 hour
		Practical Exercise (knowledge analysis and modelling using software tools)	2 hours
Day 2	Morning	Knowledge Capture Techniques	2.5 hours
		Practical Exercise (designing and running a semi-structured interview)	1.5 hours
	Afternoon	Practical Exercise: Part 1 of a 1-day KA project	3 hours
Day 3	Morning	Practical Exercise: Part 2 of the 1-day KA project	4 hours
	Afternoon	Methodologies and Procedures (including Scoping and Project Management)	3 hours

Such a course should be supplemented with on-going support when the novice knowledge engineers take on their first KA projects.

7.4.2 Supporting Knowledge Engineers

The most important aspect when supporting knowledge engineers is to be able to overcome the typical problems that can happen during a KA project. Some of the main problems and the ways in which these can be prevented or mitigated are shown below.

- **Problem**: People agree that the project is a good idea, but it is not given the priority to be allocated the resources to succeed. **Solution**: The initiation and planning of a project must be carried out carefully and not skipped or skimped. The people that can resource the project should be fully involved and should give their full backing to the project.
- **Problem**: The knowledge engineer lacks the skills, motivation or support to do a good job. **Solution**: The knowledge engineer should have the training and information to hand that explains each step of the KA process. A support system should be in place, including project monitoring and advisory sessions. The right software tools should be available to the knowledge engineer.
- **Problem**: The knowledge engineer is too set in his/her ways of working and does not see the need to follow good KA practices. **Solution**: The knowledge engineer should understand the methodology and every step should be clearly explained including the reasons why the step is performed. The KA process should be based on good practices and lessons learnt from previous projects.
- **Problem**: KA is not the answer to the task at hand. **Solution**: Initiating a project must be carried out carefully with a full understanding of what KA can and cannot do. Projects that do not fit the mould or seem too problematical should be nipped in the bud.
- **Problem**: During the project, facts are uncovered or changes occur so that the original project remit becomes inappropriate. **Solution**: There should be defined stages in a project at which the project is reviewed and at which relevant changes can be made to the aims, scope and plan.
- **Problem**: The deliverable is fine but it does not meet the expectations of the project manager and the end-users. **Solution**: Everyone involved in the project should know what the end-product will look like throughout the project (*e.g.* during the scoping, planning and reviewing stages). The project should not be oversold as this will lead to unreasonably high expectations.
- **Problem**: The project goes well and creates a good, knowledge-rich end-product, but it is not used to its full potential. **Solution**: End-users should be consulted and informed at key stages of the project. It is essential that they are involved in scoping and in defining the content and operation of the end-product. The end-product should be well-packaged, well-publicised, easy to access and provide clear benefits to the organisation.

I have tried to include many measures in the 47-step procedure that cover these points and so try to avoid the problems that can affect a KA project.

7.4.3 Dealing with Questions, Worries and Criticisms

In supporting knowledge engineers, the programme manager and support team must be able to address the questions, worries and criticisms that can be made regarding the knowledge programme and specific KA projects. Some of the typical negative 'schools of thought' that people (normally outside the knowledge programme) can have are as follows.

7.4.3.1 The 'Knowledge is Power' School of Thought
Some people feel that experts will not want to pass on their knowledge because this will give away their 'trade secrets' and reduce their position in the organisation. In practice, this is a very rare situation, as the vast majority of experts are happy to talk about their expertise (and show off how much they know!). When this problem does occur, it is important that the person concerned understands what the knowledge is to be used for, and that any worries are discussed and resolved. This brings me to an important issue: that people in an organisation should feel that they are part of a team and that sharing knowledge is a good thing for the organisation. HR has an important role to play here in establishing a culture of knowledge sharing, and ensuring that people are supported and rewarded when they do share knowledge.

7.4.3.2 The 'Simple is Best' School of Thought
Some people believe that knowledge can be acquired and shared without all of the complexity of a KA project. It certainly is the case that some knowledge can be passed from one person to another person (as has been the traditional method in many crafts and trades). But if you want to do serious things with knowledge that can maximise its use in a 21st Century organisation and remove many hidden inefficiencies and mistakes, then a well-run KA project is the answer. If anyone questions the need to use multiple methods and techniques within the KA project, then there is a nice quotation from psychologist Abraham Maslow that you might like to use: "If the only tool you have is a hammer, you tend to see every problem as a nail". And if people doubt the complexities of the human mind, then try this mind-bending quote from physicist Emerson M. Pugh: "If the human brain were so simple that we could understand it, we would be so simple that we couldn't."

7.4.3.3 The 'Knowledge Cannot be Captured' School of Thought
Some people believe that knowledge cannot be captured and that the end-product of a KA project contains information rather than knowledge. This often stems from people having a restricted or misconceived understanding of 'knowledge', and from endless debates on the differences between data, information and knowledge. People even use substitute words such as 'know-how' and 'wisdom' instead of just saying 'knowledge'. My advice to every knowledge engineer is simple: do not worry about these word games. Always ask yourself this question: "Is what I'm acquiring going to help my end-users (whether human or computer) to think and

behave more like an expert?". If yes, then that is all that matters (and if it is yes, I would say that you do have 'knowledge' in your end-product).

7.4.3.4 The 'This is Not for Us' School of Thought

Some people believe that knowledge acquisition, knowledge engineering and knowledge management are not relevant for their parts of an organisation. This may be true in some instances, but it is rare since most parts of an organisation can benefit from one or all of these areas. The potential for knowledge initiatives to affect and benefit all parts of an organisation is likely to increase markedly over the next few years. This stems from the need for organisations to remain competitive in an increasingly global market. This is evident in a recent survey by the Economist Intelligence Unit (EIU, 2006) that asked 1650 executives from around the world for their views on how their companies, and the environment in which they operate, would change over the next 15 years. When asked for the area of activity that offers the greatest potential for productivity gains over the next 15 years, the top response was "Knowledge Management".

7.5 Closing Remarks

I started this book by saying that knowledge acquisition is easy to do badly and difficult to do well. What has struck me in writing this book is just how much knowledge I have accumulated over the years and how many things have to be considered to achieve a successful KA project. Writing this book has been something of a KA project in itself, with me trying to elicit knowledge from myself and present it in a clear, well-structured and useful way. I hope I have succeeded in this.

The more I have worked in the world of knowledge, the more I have become fascinated and intrigued by its complexities, challenges and potential. I have come to hold a genuine belief that knowledge has a vital role to play in three of the most important things in our lives: **people**, **computers** and **organisations**. Unlike almost anything else, knowledge has the power to enable these three things to communicate and do things among themselves and between each other that offers tremendous possibilities for the future.

I would not say 'Knowledge is power', but rather I would say 'Knowledge is powerful'. It is powerful in many areas of our lives, and it is likely to become even more powerful as we learn more about the acquisition, storage, sharing and use of it.

Exciting times are ahead for those working in the world of knowledge. I hope this book can play a part in its future.

Appendices

Appendix 1: General Technological Ontology (GTO)

GTO (General Technological Ontology) is a generic ontology that I developed as the starting point for k-bases in any project in a technological organisation.

The aims of GTO are: (i) To standardise k-bases across an organisation, thus making it easier to see what other people have done and easier for the support team; (ii) To make knowledge analysis and modelling more efficient by reducing the time taken to set-up a generic ontology and meta-model for the k-base; (iii) To prompt for knowledge gaps during elicitation; (iv) To improve the training course and make things easier for knowledge engineers; (v) To provide a standard format for end-products and enterprise-wide k-bases.

GTO specifies 17 concepts that are:

- **People**: The names of people involved in the domain;
- **Roles**: The jobs and responsibilities of people involved in the domain;
- **Org Units**: Organisations and parts of organisations;
- **Knowledge Areas**: The areas of knowledge involved in the project;
- **Info Resources**: Resources that hold information relevant to the project;
- **Software**: Software programs and IT systems;
- **Phys Entities**: Physical things and materials;
- **Phys Phenomena**: The ways physical entities interact with each other;
- **Mental Concepts**: Non-physical (abstract) things;
- **Locations**: The places in which things are located;
- **Functions**: The purpose of a product and/or each of its sub-components;
- **Tasks**: Activities performed to satisfy a goal or set of objectives;
- **Decision Points**: Points in a flow of tasks in which a decision is made;
- **Events**: Happenings that occur in the domain;
- **Triggers**: Special events that initiate tasks in the domain;
- **Examples**: Specific illustrations of knowledge in the domain;
- **Advice**: Important pieces of knowledge that end-users should know

GTO also specifies 22 relations that provide associations between pairs of concept classes. GTO also provides several generic k-models such as concept tree, matrices, concept map and process map.

For more information on GTO, see http://www.pcpack.co.uk/GTO

Appendix 2: Effort Required for Knowledge Acquisition

The knowledge engineering effort required for knowledge acquisition (interviews, analysis and modelling) varies according to two main factors. The first is the percentage of the expert's knowledge that is to be acquired. For example, if the expert has a detailed understanding of 5 areas or of 5 main tasks, then a project tackling any one of these areas/tasks would be 20%, a project tackling two areas would be 40%. The second factor is the level of detail of the acquired knowledge, *i.e.* the depth to which the knowledge is acquired and the thoroughness of the content in the end-product. There are three levels that we can consider:

- Level 1: Basic knowledge; Essential advice; Outline of tasks; Consideration of typical situations, problems and issues.
- Level 2: Detailed knowledge; Full description of major tasks; Consideration of all major situations, problems and issues.
- Level 3: Very detailed knowledge; Comprehensive description of all tasks; Thorough consideration of all situations, problems and issues.

Table A2.1 shows the knowledge engineering effort required based on the two factors just described.

Table A2.1. Knowledge engineering effort required for knowledge acquisition

Percentage of the expert's knowledge	Level of detail of the acquired knowledge		
	Level 1	Level 2	Level 3
10%	1 wk	2 wks	3 wks
20%	2 wks	4 wks	6 wks
40%	4 wks	8 wks	12 wks
80%	8 wks	16 wks	24 wks

The time required from the expert for interviews and validation during the detailed knowledge acquisition phase, is shown in Table A2.2.

Table A2.2. Time required from the domain expert for interviews and validation

Percentage of the expert's knowledge	Level of detail of the acquired knowledge		
	Level 1	Level 2	Level 3
10%	0.5 day	1 day	1.5 days
20%	1 day	2 days	3 days
40%	2 days	4 days	6 days
80%	4 days	8 days	12 days

Glossary

AI (Artificial Intelligence). The discipline of building special computer systems that can perform complex activities usually only performed by humans.

Attribute. A quality or characteristic, *e.g.* weight, colour, usefulness.

Concepts. The individual items in a knowledge base, that represent things such as physical entities, people, tasks, issues, documents, *etc.*

Conceptual Knowledge. That part of expertise associated with the properties of concepts and the relationships between concepts. Also called 'declarative knowledge'.

Consensus Session. A meeting held between a knowledge engineer and experts to resolve and/or document differences of opinion between experts.

Decision Support System. An AI system that provides advice and suggestions to a human user.

Domain. The subject area that a KA project is focused upon.

End-product. The deliverable from a KA project, such as a knowledge document, a knowledge web or an ontology.

End-users. The people or computer systems that use an end-product.

Expert. A person with substantial experience and expertise in a particular area, who is often the main source of knowledge for a KA project. Also called 'domain expert' or 'subject matter expert'.

Expert System. An AI system that emulates the problem-solving capabilities of a human expert.

K-base (Knowledge Base). A special database that holds information representing the expertise of a particular domain.

KBE (Knowledge Based Engineering). The activity of creating software applications that incorporate the expertise of design engineers.

KBS (Knowledge Based System). An AI system that is filled with knowledge.

K-models (Knowledge models). Views of a knowledge base using diagrams and other structured representations, such as trees, maps, matrices and k-pages.

Knowledge Acquisition (KA). The activity of capturing, structuring and representing knowledge from any source for the purpose of storing, sharing or implementing the knowledge.

Knowledge Document. A document delivered by some KA projects that details the knowledge required to be coded into an intelligent software system, such as an expert system, a KBS or a KBE application.

Knowledge Elicitation. The activity of capturing knowledge from a human expert.

Knowledge Engineer. The role of a person within a KA project who performs most of the work, *i.e.* scopes and plans the project, interviews the experts, creates k-models and transforms the k-base into a useful end-product.

Knowledge Objects. The elements that make-up a k-base, *i.e.* concepts, relations, attributes and values.

Knowledge Programme. An initiative in an organisation that involves all of the KA activities and other knowledge-based activities.

Knowledge Support Team. A group of people that has experience in all aspects of KA whose roles include training and supporting knowledge engineers.

K-page. A structured page of text, images and hyperlinks that shows what is contained in the k-base about a particular concept. Also called 'annotation page'.

Meta-model. A concept map showing all the main classes of concepts and relationships between them. Used for setting up a k-base ontology and templates.

Ontology. A structured representation or store of information that describes a body of knowledge and is often encoded in special computer formats (such as OWL).

OWL (Web Ontology Language). An XML format that adds meaning to the content of web files so that computer programs can read/use web-based resources.

Procedural Knowledge. The expertise required by a person or group of people to perform a complex process, task or activity.

Project Description Scheme. A framework for classifying and/or describing a particular project (*e.g.* by its properties, aims, deliverables). Used for planning and decision-making during a KA project.

Project Team. The people who perform activities on a KA project.

Role Sheet. A page of information describing the responsibilities and activities required of a specific role during a KA project (*e.g.* Expert Role Sheet, End User Role Sheet, Project Manager Role Sheet).

Scoping. The activity of selecting the specific areas of knowledge to be acquired during a KA project.

Triple. A relationship between two concepts, *e.g.* 'book – written by – author'. So called, because there are three elements to the expression.

Value. A specific quality or characteristic of a concept, *e.g.* heavy, red, useful.

XML. A format for representing information in a web file. For example, rather than just have the word 'car' in a web file, the XML could hold extra (hidden) information such as: car is a vehicle, and car has the synonym 'automobile'.

References and Bibliography

References

Booch G, Rumbaugh J, Jacobson I (2005) The Unified Modeling Language User Guide, 2nd Edition. Addison Wesley, Reading, MA.

Boose JH (1988) Uses of Repertory Grid-Centered Knowledge Acquisition Tools for Knowledge-Based Systems. International Journal of Man-Machine Studies 29:287-310

Cañas AJ, Novak JD, González FM (eds.) (2004) Proceedings of the First International Conference on Concept Mapping. Pamplona, Spain 2004.

EIU (2006) Forefront 20/20: Economic, Industry and Corporate Trends. Economist Intelligence Unit, London <http://www.eiu.com/ >

Ericsson KA, Simon HA (1980) Verbal reports as data. Psychological Review 87:215-251

Fensel D (2004) Ontologies: A Silver Bullet for Knowledge Management and Electronic Commerce (Second Edition). Springer-Verlag, Berlin Heidelberg New York.

Gaines BR, Shaw MLG (1993) Basing Knowledge Acquisition Tools in Personal Construct Psychology. Knowledge Engineering Review 8:49-85

Hoffman RR (1987) The Problem of Extracting the Knowledge of Experts from the Perspective of Experimental Psychology. AI Magazine 8:53-67

Klein GA (1996) The Development of Knowledge Elicitation Methods for Capturing Military Expertise. U. S. Army Research Institute for the Behavioral and Social Sciences <http://mentalmodels.mitre.org/cog_eng/reference_documents/>

Rugg G, McGeorge P (1997) The Sorting Techniques: a tutorial paper on card sorts, picture sorts and item sorts. Expert Systems 14:80-93

Schreiber ATh, Akkermans JM, Anjewierden A, De Hoog R, Shadbolt N, Van De Velde W, Wielinga, B (2000) Knowledge Engineering and Management: The CommonKADS Methodology. MIT Press, Cambridge, MA.

Shadbolt NR (2005) Eliciting Expertise. In: Wilson JR, Corlett EN (eds.) Evaluation of Human Work, 3rd Edition. Taylor and Francis, London.

Stokes M (ed.) (2001) Managing Engineering Knowledge: MOKA Methodology for Knowledge Based Engineering Applications. Professional Engineering Publishing Ltd., London and Bury St. Edmonds.

Zaff B, McNeese M, Snyder D (1993) Capturing Multiple Perspectives: A User-Centred Approach to Knowledge and Design Acquisition. Knowledge Acquisition 5:79-116.

Bibliography

Knowledge Elicitation and Acquisition

Boose JH (1989) A Survey of Knowledge Acquisition Techniques and Tools. Knowledge Acquisition 1:39-58

Cooke NJ (1994) Varieties of Knowledge Elicitation Techniques. International Journal of Human-Computer Studies 41:801-849

Corbridge C, Rugg G, Major N, Shadbolt N, Burton AM (1994) Laddering: Technique and Tool Use in Knowledge Acquisition. Journal of Knowledge Acquisition 6:315-341

Diaper D (ed) (1989) Knowledge Elicitation: Principles, techniques and applications. EllisHorwood Ltd, Chichester, UK.

Hart A (1986) Knowledge Acquisition for Expert Systems, McGraw-Hill, Inc., New York, NY.

Hoffman RR, Shadbolt NR, Burton AM, Klein GA (1995) Eliciting Knowledge from Experts: A Methodological Analysis. Organizational Behavior and Decision Processes 62:129-158

Milton N, Shadbolt N, Cottam H, Hammersley M (1999) Towards a Knowledge Technology for Knowledge Management. International Journal of Human-Computer Studies 51:615-641.

Knowledge Management

Davenport TH, Prusak L (1998) Working Knowledge: How Organisations Manage What They Know. Harvard Business School Press, Boston, MA.

Rumizen MC (2002) The Complete Idiot's Guide to Knowledge Management. CWL Publishing Enterprises, Madison, WI.

Shadbolt N, Milton N (1999) From Knowledge Engineering to Knowledge Management. British Journal of Management 10:309-322.

Tacit Connexions website: www.tacitconnexions.com

Psychology

Chi MH, Glaser R, Farr M (eds.) (1988) The Nature of Expertise. Lawrence Erlbaum Associates, Hillsdale, NJ.

Jonassen D, Beissner K, Yacci M (1993) Structural Knowledge: techniques for representing, conveying and acquiring structural knowledge. Lawrence Erlbaum Associates, Hillsdale, NJ.

Smith JA, Harré R, Van Langenhove L (eds.) (1995) Rethinking Methods in Psychology. Sage Publications, London.

Milton N, Clarke D, Shadbolt N (2006) Knowledge Engineering and Psychology: Towards a Closer Relationship. International Journal of Human-Computer Studies 64:1214-1229.

Software Tools

CmapTools: http://cmap.ihmc.us/
PCPACK: http://www.epistemics.co.uk
Protégé: http://protege.stanford.edu/
Repertory Grid tools: http://repgrid.com/
UML tools: http://www-306.ibm.com/software/rational/

Index